A Guide to Particulate Science in Pharmaceutical Product Development

The pharmaceutical applications of powder technology have long been recognized. Yet while many books focus on aspects of powder formation and behavior, there are few texts that explore the power of particulate science in the design, manufacture, and control of quality medicines. This revision discusses key principles and practical applications. The authors cover particulate material, its form and production, sampling from bodies of powder, particle size descriptors and statistics, behavior of particles and powder, instrumental analysis, particle size measurement and synergy of adopted techniques, and in vitro and in vivo performance criteria. Case studies are included in this new edition.

This fully revised edition:

- Provides an essential account of particulate science including several new chapters on multicomponent particles, regulatory considerations and product development.
- Presents a variety of topics ranging from the quality of published data on particle size in pharmaceuticals to the future of crystal engineering.
- Reviews methods of particle measurement and their importance for specific applications.
- Discusses misconceptions and misunderstandings of particulate science together with lessons from other industries.

Drugs and the Pharmaceutical Sciences

A Series of Textbooks and Monographs

Series Editor
Anthony J. Hickey
RTI International, Research Triangle Park, USA

The Drugs and the Pharmaceutical Sciences series is designed to enable pharmaceutical scientists to stay abreast of the changing trends, advances, and innovations associated with therapeutic drugs and that area of expertise and interest that has come to be known as the pharmaceutical sciences. The body of knowledge that those working in the pharmaceutical environment have to work with, and master, has been, and continues, to expand at a rapid pace as new scientific approaches, technologies, instrumentations, clinical advances, economic factors, and social needs arise and influence the discovery, development, manufacture, commercialization, and clinical use of new agents and devices.

Recent Titles in Series

Good Design Practices for GMP Pharmaceutical Facilities, Second Edition, Terry Jacobs and Andrew A. Signore

Handbook of Bioequivalence Testing, Second Edition, Sarfaraz K. Niazi

FDA Good Laboratory Practice Requirements, First Edition, Graham Bunn

Continuous Pharmaceutical Processing and Process Analytical Technology, Ajit Narang and Atul Dubey

Project Management for Drug Developers, Joseph P. Stalder

Emerging Drug Delivery and Biomedical Engineering Technologies: Transforming Therapy, Dimitrios Lamprou

RNA-seq in Drug Discovery and Development, Feng Cheng and Robert Morris

Patient Safety in Developing Countries: Education, Research, Case Studies, Yaser Al-Worafi

Industrial Hygiene in the Pharmaceutical and Consumer Healthcare Industries, Casey Cosner

Cancer Targeting Therapies: Conventional and Advanced Perspectives, Muhammad Yasir Ali, Shazia Bukhari

Molecular Recognition in Pharmacology, Mikhail Darkhovskiy

GMP Audits in Pharmaceutical and Biotechnology Industries, Mustafa Edik

Purification of Biotechnological Products: A Focus on Industrial Applications, Adalberto Pessoa Jr, Beatriz Vahan Kilikian and Paul Long

Principles of Research Methodology and Ethics in Pharmaceutical Sciences: An Application Guide for Students and Researchers, Vikas Anand Saharan, Hitesh Kulhari and Hemant Jadhav

Good Clinical Practices in Pharmaceuticals, Graham P. Bunn.

An Introduction to Generative Drug Discovery, Sean Ekins

Viral Oncology: New Approaches to Molecular Cancer Therapeutics, Edited By Umesh Kumar, Deepak Parashar, Subodh Kumar

A Guide to Particulate Science in Pharmaceutical Product Development, Margaret Louey, Timothy Crowder and Anthony J. Hickey

For more information about this series, please visit: www.crcpress.com/Drugs-and-the-Pharmaceutical-Sciences/book-series/IHCDRUPHASCI

A Guide to Particulate Science in Pharmaceutical Product Development

Margaret D. Louey,
Timothy M. Crowder, and Anthony J. Hickey

CRC Press
Taylor & Francis Group
Boca Raton London New York

CRC Press is an imprint of the
Taylor & Francis Group, an **informa** business

Second edition published 2026
by CRC Press
2385 NW Executive Center Drive, Suite 320, Boca Raton, FL 33431

and by CRC Press
4 Park Square, Milton Park, Abingdon, Oxon, OX14 4RN

CRC Press is an imprint of Taylor & Francis Group, LLC

ISBN: 978-1-032-73182-7 (hbk)
ISBN: 978-1-032-74627-2 (pbk)
ISBN: 978-1-003-47015-1 (ebk)

DOI 10.1201/9781003470151

Typeset in Times
by Deanta Global Publishing Services, Chennai, India

Contents

About the Authors

Margaret Louey is a pharmaceutical scientist with extensive experience in drug development and global health initiatives. She earned her PhD from Monash University in Melbourne, Australia, and then completed a post-doctoral fellowship at the University of North Carolina in Chapel Hill. Margaret entered the pharmaceutical industry and worked at GlaxoSmithKline for 15 years, primarily in product development of respiratory products. During this time, her interest evolved to increasing access to medicines in low- and middle-income countries. Margaret transitioned to the non-profit sector and joined the Clinton Health Access Initiative, where she has supported public–private partnerships to accelerate product development and regulatory submissions for high-priority pediatric products for the treatment and prevention of HIV and tuberculosis in low- and middle-income countries.

Timothy "Timm" M. Crowder, PhD, is a pharmaceutical scientist and entrepreneur with deep expertise in particulate science and drug delivery, particularly in inhalation technologies. He earned his PhD from the University of North Carolina at Chapel Hill, where his research focused on the design and development of innovative drug delivery systems. This work sparked a lifelong interest in the complex interplay between particle engineering, device design, and clinical outcomes.

Timm's career has spanned both industry and entrepreneurial ventures, with a particular focus on transforming cutting-edge research into practical, patient-centric therapies. As a co-founder of Oriel Therapeutics, Timm helped pioneer new approaches to dry powder inhaler technology. He went on to hold leadership roles across both large pharmaceutical companies and biotech startups, where he applied his expertise to advance inhaled therapies for respiratory diseases, rare conditions, and systemic delivery.

Throughout his career, Timm has remained passionate about bridging the gap between fundamental science and real-world product development. His work has encompassed everything from dry powder vaccine formulations to advanced inhaler technologies, always with an emphasis on delivering therapies that improve patient outcomes. In addition to his industry contributions, Timm has maintained a strong connection to academic research and training, co-authoring key publications on pharmaceutical particulate science and mentoring the next generation of scientists and innovators in the field.

Anthony J. Hickey is Distinguished Fellow Emeritus at RTI International, Research Triangle Park, NC; Professor Emeritus of Pharmacoengineering and Molecular Pharmaceutics of the Eshelman School of Pharmacy; and Adjunct Professor of Biomedical Engineering of the School of Medicine at the University of North Carolina at Chapel Hill. He is Fellow of the American Association for the Advancement of Science, the American Association of Pharmaceutical Scientists, the Academy of Pharmaceutical Sciences of Great Britain, the Royal Society of

Biology, the Royal Society of Medicine, and the National Academy of Inventors. He received the Research Achievement Award of the Particulate Presentations and Design Division of the Powder Technology Society of Japan, the David W Grant Award in Physical Pharmacy of the American Association of Pharmaceutical Scientists, Thomas T Mercer Joint Prize for Excellence in Inhaled Medicines and Pharmaceutical Aerosols of the American Association for Aerosol Research and the International Society for Aerosols in Medicine, and the Ralph Shangraw Memorial Award for Excipient and Excipient Technology of the International Pharmaceutical Excipient Consortium Foundation. He is Founder (and formerly President and CEO) of Cirrus Pharmaceuticals, Inc.; Founder (formerly CSO) of Oriel Therapeutics, Inc; CEO of Astartein, LLC.; and Member of the Board of First Flight Venture Center, Inspiring Pty Ltd., and KAER Biotherapeutics. He also serves as Chair of the United States Pharmacopeia (USP) Joint Sub-Committee on Nanotechnology and as Member of the USP Pharmaceutical Dosage Forms Expert Committee (DFEC) and Inhalation Sub-Committee of the DFEC of the United States Pharmacopeia. He consults on multidisciplinary research programs in the field of pulmonary drug and vaccine delivery for the treatment and prevention of a variety of diseases and serves on numerous editorial and scientific advisory boards.

Preface

The original version of this volume, published in 2003, included a dedication to the co-author, Dr Norman Orr, who unfortunately passed away before it appeared in print. His experience in the pharmaceutical industry over several decades had led to his holding very strong opinions about the role of pharmaceutical particulate science and technology in the quality and performance of drug products. His vision of the need for a text on this topic was captured in the following statement.

Powder technology is the foundation of dosage form design and particle engineering is the future of efficient, reproducible, and effective drug delivery.

He clearly had a premonition of the extent to which novel technologies, many of which have emerged in the new millennium, would rely on an understanding of particles, their physicochemical characteristics, and their role in the quality and performance of the dosage form. His appeal to the community of pharmaceutical scientists to give greater attention to particulate science and technology has long since been met. Some of the criticism leveled at the paucity of data generated on particulate components of pharmaceutical dosage forms in the first edition is no longer relevant and has not been reproduced in the present edition.

This book does not simply describe particle and powder properties and methods of measurement. It is unique in presenting a critique of the utility of techniques and data collected, the context in which they are used, and the strengths and weaknesses of interpretation with respect to both product development and regulatory needs. This innovative approach makes it an ideal complement to fundamental and comprehensive, general, and pharmaceutical texts describing particles and powders. By guiding the reader through critical considerations of solid materials at a small scale with respect to dosage form optimization, a greater understanding of the interpretation and application nuances can be achieved. The hope is to serve the goal of minimizing empiricism in manufacture and characterization and maximizing the return on investment with respect to any experiments or practical work required in pharmaceutical product development.

In this edition, the contents have been restructured to follow the product development and regulatory pathway sequentially for clarity to the reader. The rationale for this change reflects important advances since the first volume was published. At that time, the field was on the cusp of a major shift in quality control and assurance, a significant component of which was identified in the US FDA guidance on product development, describing principles of quality-by-design, promulgated in 2009. Commentary in the original volume advocated for rigorous monitoring and control of particles and powders with the tools and instruments that were then available but not used consistently. The need for detailed and relevant assessment continues to be required to assure the quality of the product. However, an acerbic text calling on the pharmaceutical scientist to attend to these issues is no longer necessary. The adoption of continuous rather than batch processes and the ability to monitor and control manufacturing in real time has brought quality-by-design principles into the forefront of product development. Consequently, this edition serves not to focus the scientific community on past shortcomings but as a high-level overview to be consulted prospectively with other sources, including those online, either as an introduction or as a reference text depending on the education and training of the reader.

We believe this volume will be a valuable addition to the Drugs in the Pharmaceutical Sciences series and are grateful for the opportunity to expand and update the original text.

Margaret D. Louey,
Chapel Hill, NC
Timothy M. Crowder,
Durham, NC
Anthony J. Hickey,
Chapel Hill, NC

Acknowledgments

Dr. Norman Orr was a thought leader in pharmaceutical particle science and contributed substantially to the first edition of this book. Unfortunately, he passed away before it was published and missed the major developments of the last 20 years. Many of his concerns regarding the need for a more thorough investigation of particle properties and their role in the quality and performance of drug products have since been addressed. He would be pleased with the progress that has been made. We like to think that his professional contributions placed him in the vanguard of those moving the field towards a quality-by-design paradigm at the turn of the millennium.

Dr. Anthony J Hickey's mentorship and guidance have had a profound impact on Margaret and Timm from our graduate studies and throughout our professional careers. Tony's insatiable curiosity, interdisciplinary approach, and adaptability across various fields have truly inspired us to continue learning, growing, and applying our expertise toward improving human health. We thank Tony for providing us the opportunity to co-author the first edition of this book with our deep fundamental knowledge honed in his lab, and now again for the second edition with wider practical experience to offer.

1 Introduction and Overview

The first edition of this book sought to address specific questions that the authors believed were often overlooked in standard texts on particle technology and characterization. At the time of its publication, it was one of the very few books focused on pharmaceutical particulate systems. Consequently, it drew largely on primary literature in the pharmaceutical sciences and more general text in powder and particle science and technology.

Since the original publication, changes in the ease of access to primary literature and the pace and scale of publication of new research have reduced the relevance of a book-format broad survey to scientists and engineers who wish to understand particular issues in particulate science. In addition, it is no longer the case that those with a background in relevant fields are the only researchers who might be interested in familiarizing themselves with the literature on this topic. Consequently, the need that the authors hope the present volume fulfills differs from the original text.

To provide new relevance, the book has been completely restructured to cover topics in a sequential manner with respect to the development of drug products. The new arrangement of content falls into broad categories that reflect: (1) the solid state and its manipulation through the processing of drug and excipient, (2) the preparation of powder constructs from source particulate materials, and (3) the use of the optimized powder construct in pharmaceutical dosage forms. Commentary is given on the monitoring and controls applied at each stage to ensure product quality in order to support accurate and reproducible dose delivery, which is essential for the safety and efficacy of the drug in disease therapy. Figure 1.1 illustrates the sequence in which the book is organized to link specific aspects of particle science and technology to the objective of product development and clinical application.

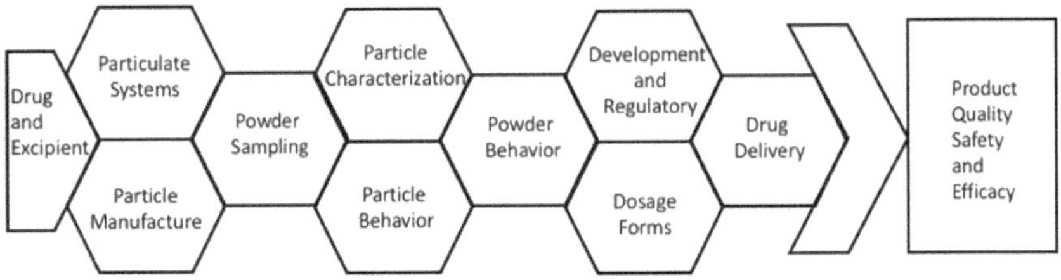

FIGURE 1.1 Schematic depiction of the sequence in which particles and powders may be considered with respect to pharmaceutical product development.

DOI: 10.1201/9781003470151-1

PHARMACEUTICAL PRODUCTS

The pharmaceutical reality is that in 90 percent of all medicines, the active ingredient is in the form of solid particles. Although the general population has a clear visual appreciation of particulate systems, such as sand, gravel, stones, rocks, and boulders, and is capable of differentiating qualitatively among these, the ability to uniquely define particulate systems in a manner that suits all purposes is elusive.

Particle size provides the basis for building quality into and optimizing the design of medicine, but it is seldom fully exploited. Raw materials are very inconsistent, and how this affects the activity of the medicines is complex.

Significant opportunities surround the engineering of surfaces and morphology of crystals. The implications for formulation design and batch-to-batch uniformity are immense. This is leading to greater consistency in secondary manufacture. Ultimately, this approach will result in cheaper, well-designed, and higher-quality medicines. In addition, the move from batch to continuous process that allows real-time monitoring and control of key variables based on knowledge of particulate science and technology will allow the preparation of powder and construction of dosage forms within the specifications required to meet quality standards at reduced efficiency and at reduced cost.

INFLUENCE OF PARTICLE PROPERTIES ON PHARMACEUTICAL QUALITY

The process control and analytical monitoring of particle properties underpin the quality of solid drug dosage forms which, in turn, assures the accuracy and reproducibility of performance required to establish safety and efficacy.

The following chapters review the literature with respect to the relationship of particle size to pharmaceutical parameters. The rigor in the design of experiments is discussed, especially related to the following: control of input variables includes fully characterizing the components of a formulation and controlling processing variables within limits that facilitate uniformity of drug distribution and the dose delivered by a dosage form. The tools that are employed to characterize the product include: preparation and characterization of known particle size fractions, representative sampling, and indirect methods of characterization. It is important to recognize that the methods of characterizing the physicochemical properties and performance of particles are complementary and create a body of evidence to support the uniformity between batches during manufacture and individual dosage forms. Differences in particle size are a primary source of variability in drug products. Consequently, the complementary nature of techniques can be illustrated by considering particle sizing methods.

POWERFUL METHODS COMPLEMENTARY TO PARTICLE SIZING

A mathematical interpretation of bulk properties such as the flow of powders, coalescence of emulsions, and bleeding in ointments can be developed by using approaches in complexity analysis as well as stochastic and statistical phenomena. Experimental investigations on the powder flow of pharmaceutical systems, including raw materials, excipients, intermediate granulations, and lubricated granulation for encapsulation and tableting processes, should provide numerical values that relate to features of the component particles. These features may be simple, such as size, or more complex, such as a combination of size, shape, asperities, interparticle forces, and environment in interparticle spaces.

Comparison of such data with that derived from the application of such conventional techniques as sieving may enhance our understanding of the fundamental interactions leading to successful granulation. A broad view of how these data and their interpretation fit into pharmaceutical development embraces a wide range of concepts, from mathematical descriptors of particle shape and molecular probes of surface energetics to numerical definitions of bulk properties. Such methods enable revolutionary approaches to understanding processes such as disintegration, which in certain biological circumstances is a key precursor to dissolution.

Particulate systems may be described, or evaluated, in three contrasting but complementary ways:

1. Direct phenomenological characterization (e.g., bulk flow and shear cell);
2. Knowledge of component particles, their surfaces, the microenvironment in which they exist, and theoretical models that support their interpretation;
3. Mathematical interpretation of discontinuities resulting in the observed phenomena (e.g., coalescence of emulsions, avalanching powder flow, bleeding in ointments, and disintegration of tablets, or the sudden disruption of the tablet from the bulk). Such knowledge may translate into an improved product.

The ability to integrate these activities into computer-based models that allow for rapid translation into product development has been a major advance. It is now possible to identify specific critical quality attributes and assign monitoring and control specifications such that the quality can be assured without recourse to independent studies.

INTEGRATION OF PARTICLE SIZING METHODS WITH THEIR IN-USE SITUATION

Powder science may be thought of as the understanding of particulate materials at a molecular level, their behavior as individual particles, and the physical and mechanical properties of collections of particles under defined conditions or in specific processes. The formulation can be defined as the transformation of a new chemical entity (NCE) into a medicine that is convenient to manufacture, distribute, and use, such that, upon administration, the NCE is delivered in a predictable, qualified, and desirable manner. For the majority of medicines from the process of discovery/

isolation of the NCE through production as a medicine until dissolution (e.g., in the gastrointestinal tract or GIT), the NCE is in particulate form.

The nature of this particulate form (morphology, chemistry of the surface, and physical size) varies, depending upon its place in the transformation process. At any given time, the form of a particle may be critical to key parameters. For example, during secondary production, the shape, particle size, and surface energetics may be critical for successful granulation and subsequent compression; the solid-state stability of a capsule or tablet may well depend upon size morphology and excipient/drug particle adjacencies; and the content uniformity of potent drugs depends upon both the mass of the individual particles and the extent to which these individual particles are adequately dispersed and not agglomerated.

A desirable feature of any particulate sizing program for a given NCE in a dosage form would be to monitor the size and characteristics of that particle during the entire process, from isolation during primary manufacture to disintegration in the GIT. This, then, is the extent of the challenge. While there remains no question that troubleshooting may still be required to investigate deviations in product quality and performance. Current practices in the pharmaceutical industry are at the cutting edge of engineering and information technology, leaving very little outside the boundaries of consideration.

ADJACENCIES AND INTERACTIONS BETWEEN PARTICLES

Particle–particle interactions do occur, and each particle must be considered in the context of its environment. Adjacencies and interactions between particles fall into two categories: advantageous and disadvantageous. The type of interactions being sought will depend upon the aspect of the design or process being considered.

As an example, when considering content uniformity and dissolution, the decision may be that particles should be <10 μm in size. These particles must be free flowing and must disperse and mix with excipients readily. If the mix is used in a direct compression process, it would be desirable that strong interaction occur between the excipient and drug but not between adjacent excipient particles (e.g., magnesium stearate when used as an excipient in tablet compression).

Technology exists to specifically assess the extent, at any processing stage, of particle adjacencies and interactions. Each of the steps is concerned with either breaking particle–particle interactions, or encouraging the formation of new particle–particle interactions, or both processes (i.e., mixing, granulation, tableting, and disintegration).

PARTICLE CHARACTERIZATION AND ITS IMPACT ON REGULATORY SUBMISSIONS FROM FORMULATION THROUGH TOXICOLOGY AND EFFICACY

The interface between the dosage form and the organism is the basis for the pharmacological or toxicological effects observed. The absolute amount of a drug or xenobiotic delivered by any route of administration and its rate of release, which will influence the bioavailability, must be considered since efficacy and toxicity are dose related. The nature of the particulate components of the drug delivery system will ultimately dictate the rate of release and instantaneous dose. The crystal structure, presence of polymorphs, and degree of subdivision all contribute to the solubility and dissolution rate of drugs. In general terms, these features influence the circulating concentrations of the agent, bringing it to therapeutic or toxic levels. However, in certain circumstances, specific organs may experience higher or lower concentrations of the drug than circulating levels based on intentional or unintentional targeting. Inhalation aerosol delivery, for example, deliberately targets the lungs and achieves locally therapeutic doses while minimizing systemic side effects. Oral delivery of drugs results in absorption to a blood supply, which is delivered to the liver by the hepatic portal vein before distribution to the rest of the body. This represents unintentional targeting of an organ, which may result in local advantageous or deleterious effects.

Minimally, a full characterization of the particulate nature of the product is required to ensure the reproducibility of dose delivery and overall quality. If the particle size and distribution are thought to directly impact the efficacy of a product, as is the case with inhalation aerosols, then this becomes a determinant of regulatory scrutiny.

CONCLUSION

The role of particulate science in the preparation of medicines is ubiquitous. The solid state is the dominant means of presentation of a drug. Once molecules have crystallized from solution or solidified from a melt, the individual form and degree of subdivision of the population must be characterized. These features directly relate to the performance of the dosage form and ultimately the way in which molecules present at the site of absorption, action, and the target receptor. Most pharmaceutical products are not prepared as particulate systems. However, inhalation products are an exception and are frequently prepared in particulate form. The impact of the particulate nature of components of the dosage form and route of administration on therapeutic effect is of the utmost importance and is considered in the following chapters.

2 Particulate Systems
Foundational Principles

STATES OF MATTER

The three states of matter—gas, liquid, and solid—each represent a different degree of molecular mobility. Gibbs first described the nature of states of matter according to the principles of thermodynamics (1, 2). Despite developments in the fields of quantum physics and chemistry, Gibbs' observations remain a valid description of the nature of matter.

Gas molecules are in constant, vigorous, random motion according to the classical ideal gas theory of Bernoulli. Consequently, they take the shape of the container, are readily compressed, and exhibit low viscosity.

Liquids exhibit restricted random motion; the volume occupied is limited by their condensed nature resulting from intermolecular forces. Thus, liquids take the shape of a portion of the container. The liquid properties of water are within everyone's daily experience. However, to illustrate the extreme, other commonly experienced substances, such as wax, pitch, and glass, are highly viscous liquids, and not, as they appear to be, solids.

The last state of matter is highly constrained and is the result of a variety of forces. Solids are of great significance in the way we experience the world and are a dominant theme in the preparation of pharmaceuticals.

In broad terms, solids may be found in amorphous or crystalline states (3, 4). Crystallinity involves the regular arrangement of molecules or atoms into a fixed, rigid, three-dimensional pattern or lattice. Many amorphous materials exhibit some degree of crystallinity. The term "crystalline" is reserved for materials that exhibit a high degree of internal regularity, resulting in the development of definite external crystal faces.

Truly amorphous solids are similar to gases and liquids in that their physical properties when measured in any direction are the same. In consequence, they are regarded as isotropic substances. Most crystals are anisotropic in that their mechanical, electrical, magnetic, and optical properties vary according to the direction of measurement (5). Cubic crystals are a notable exception because their highly symmetrical internal arrangement renders them optically isotropic. Anisotropy is most readily detected by refractive index measurement. Polarizing lenses may be employed to study anisotropy optically.

Liquid crystals are isotropic materials that can be induced to exhibit anisotropic behavior. Bose's swarm theory explains mesophase formation in liquid crystals by the formation of ordered regions in an otherwise randomly arranged "liquid." Molecules in these systems "swarm" into alignment. This phenomenon can be classified into smectic (soap-like), nematic (thread-like), and cholesteric. The smectic mesophase is

DOI: 10.1201/9781003470151-2

characterized by the flow of liquids in thin layers over each other. Cholesteric mesophases exhibit strong optical activity. Systems consisting of organic, often aromatic, elongated molecules form anisotropic liquids.

CRYSTALLINE SOLIDS

Crystals comprise a rigid lattice of molecules, atoms, or ions. The regularity of the internal structure of this solid body results in the crystal having a characteristic shape (6, 7). The growth of most natural crystals has been restricted in one or more directions, resulting in exaggerated growth in other directions and giving rise to the so-called crystal habit (8–11).

The apparent order exhibited by crystalline solids was once thought impossible to classify. In the late eighteenth century, however, Haüy, building on the observations of Steno from the prior century, proposed the law of constant interfacial angles, which stated that angles between corresponding faces of all crystals of a given substance are constant. Therefore, crystals may vary in size and development of various faces but the interfacial angles do not vary. A single substance can crystallize in more than one structural arrangement or form, known as polymorphism, an anomaly with regard to Haüy's law. However, the law does apply to each form or polymorph. Modern techniques of X-ray crystallography enable lattice dimensions and interfacial angles to be measured, by the application of Bragg's law, with high precision on milligram quantities of crystalline powders (12–15).

CRYSTAL SYMMETRY

Classification of crystals can be considered in terms of the apparent order conveyed visually. This can be described in terms of the symmetries exhibited, either about a point (center), a line (axis), or a plane. A cube is a simple example of such classification, having a single point, 13 axes, and 9 planes of symmetry, totaling 23 elements of symmetry, as shown in Figure 2.1. An octahedron exhibits the same quantity or elements of symmetry as a cube. Therefore, these polygons are related despite their apparent difference in appearance. The octahedron can be transformed into a cube by passing through the intermediate forms of truncated cube, octahedron,

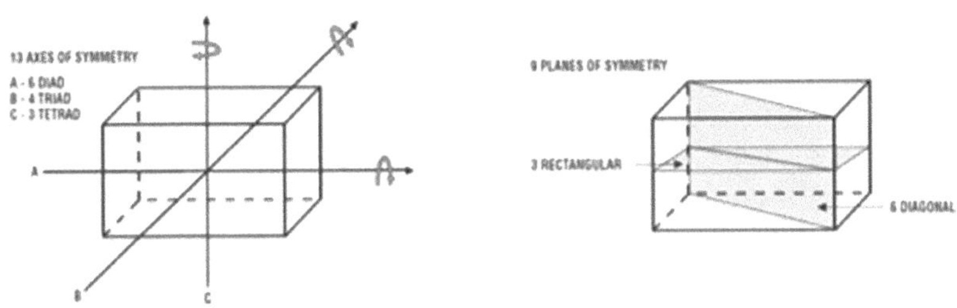

FIGURE 2.1 Classification of (a) point (center), lines (axes), and (b) planes of symmetry for a cube.

and cubooctahedron. These represent 3 of the 13 Archimedean semiregular solids, which are called combination forms. Crystals frequently exhibit combination forms. The tetrahedron is also related to the cube and octahedron. These three forms belong to the five regular solids of geometry. The other two forms, the regular dodecahedron and icosohedron, do not occur in the crystalline state. However, the rhombic dodecahedron is frequently found in nature, particularly in garnet crystals.

Euler's Relationship

Euler's relationship is particularly useful for calculating the number of faces (F), edges (E), and corners (C, vertices) of any polyhedron, according to the expression:

$$E = F + C - 2 \tag{2.1}$$

A fourth element of symmetry is known as compound or alternating symmetry, or, alternatively, symmetry about a rotation-reflection axis or axis of rotatory inversion. This type of symmetry comes about by performing two operations: rotation about an axis and reflection in a plane at right angles to the axis, as shown in Figure 2.2a. This is also called inversion about the center.

CRYSTAL SYSTEMS

Only 32 combinations of the elements of symmetry are possible. These are the point groups or classes. Most of these classes have been observed in crystalline bodies. These classes may be grouped into seven systems based on three dimensions (x, y, and z) and angles of faces (α, β, and γ). Hexagonal crystals are unique because they have six edges in two dimensions rather than four as for the other crystal systems. Hexagonal crystals are described by invoking a third dimension in two-dimensional space with a third angle (μ). Table 2.1 describes the crystal systems.

Miller Indices

An alternative visual reference technique, known as Miller indices involves using a three-dimensional Cartesian coordinate system to indicate the position of crystal faces in space. In this technique, all of the faces of a crystal can be described in terms of their axial intercepts, as illustrated in Figure 2.2b.

Space Lattices

Hooke and Haüy concluded that all crystals were composed of a large number of minute units, each shaped like the larger crystal. Consequently, a space lattice is a regular arrangement of points in three dimensions. Each of the seven crystal systems identified by elements of symmetry can be classified into 14 Bravais lattices, as shown in Table 2.2. Although these are the basic lattices, interpenetration can occur in actual crystals; this interpenetration can potentially give rise to 230 possible combinations. An alternative to the Bravais–Donnay–Harker principle considers space groups rather than lattice types. Generally, crystals cleave along lattice planes.

TABLE 2.1
Crystal Systems, the Number of Subsidiary Classes, and Crystal Structures (Dimensions x, y, and z, and Angles α, β, γ, and μ)

Crystal System	(Classes)	Angles/Dimensions
Regular	(5)	$\alpha = \beta = \gamma = 90°/x = y = z$
Tetragonal	(7)	$\alpha = \beta = \gamma = 90°/x = y \neq z$
Orthorhombic	(3)	$\alpha = \beta = \gamma = 90°/x \neq y \neq z$
Monoclinic	(3)	$\alpha = \beta = 90° \neq \gamma/x \neq y \neq z$
Triclinic	(2)	$\alpha \neq \beta \neq \gamma \neq 90°/x \neq y \neq z$
Trigonal	(5)	$\alpha = \beta = \gamma \neq 90°/x = y = z$
Hexagonal	(7)	$\alpha = \beta = \mu = 60°, \gamma = 90°$

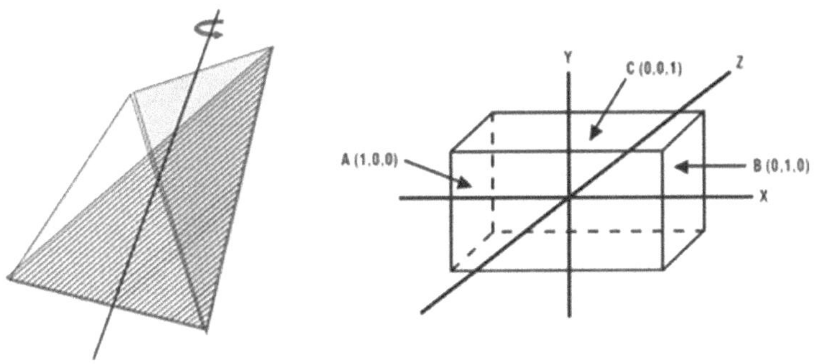

FIGURE 2.2 (a) Rotary inversion symmetry and (b) cubic geometry illustrating Miller indices.

TABLE 2.2
Fourteen Bravais Lattices and Their Equivalent Symmetries and Crystal Systems

Type of Symmetry	Lattice	Crystal System
Cubic	Cube Body centered Face centered	Regular
Tetragonal	Square prism Body centered	Tetragonal
Orthorhombic	Rectangular prism Body centered Rhombic prism Body centered	Orthorhombic
Monoclinic	Monoclinic parallelepiped Clinorhombic prism	Monoclinic
Triclinic	Triclinic parallelepiped	Triclinic
Rhomboidal	Rhombohedron	Trigonal
Hexagonal	Hexagonal prism	Hexagonal

SOLID STATE BONDING

Four types of crystalline solids may be specified: ionic, covalent, molecular, and metallic. Some materials are intermediate between these types.

Ionic crystals (e.g., sodium chloride) consist of charged ions held in place in the lattice by electrostatic forces. Each ion is separated from oppositely charged ions by regions of negligible electron density.

Covalent crystals (e.g., diamond) consist of constituent atoms, which do not carry effective charges. A framework of covalent bonds, through which their outer electrons are shared, connects these atoms.

Molecular crystals (e.g., organic compounds) are discrete molecules held together by weak attractive forces (π-bonds, hydrogen-bonds).

Metallic crystals (e.g., copper) comprise ordered arrays of identical cations. The constituent atoms share their outer electrons, which are free to move through the crystal and confer "metallic" properties on the solid.

Pharmaceutical products are largely limited to ionic and molecular crystals. The ionic crystals are associated with additives to the products; the drug substance itself is usually a molecular crystal. However, exceptions include sugars (e.g., lactose), which may be used as excipients, and salts (e.g., potassium chloride), which have physiological activity.

Isomorphs and Polymorphs

Polymorphs are different crystalline forms of the same chemical entity, which may exhibit different physicochemical properties (16). Different polymorphic forms or changes in solvation may be detected by using bulk analytical methods, including thermal analysis (such as differential scanning calorimetry [DSC] and thermogravimetric analysis [TGA]) (17–20), X-ray diffraction, or near-infrared spectroscopy (21, 22). Amorphous regions are thermodynamically unstable and exist in a higher energy state than crystalline regions within the particle (14, 23). X-ray diffraction and DSC may detect large amounts of amorphous material. Smaller amounts of amorphous phase may be detected by moisture vapor sorption, isothermal microcalorimetry, and inverse gas chromatography (14, 23).

Mitscherlich's law of isomorphism describes in general terms how valency dictates isomorph formation according to the following expression for alums:

$$M'_2SO_4 \cdot M'''_2(SO_4)_3 \cdot 24H_2O \tag{2.2}$$

where M′ is univalent (e.g., potassium or ammonium) and M‴ is trivalent (e.g., aluminum, chromium, or iron).

Many phosphates, arsenates, sulfates, and selenates are isomorphic. Isomorphous materials can often form mixed crystals by cocrystallization from solutions.

Enantiomorphs and Racemates

Enantiomorphs and racemates may be defined as two crystals of the same substance that are mirror images of each other. These crystals have neither planes of symmetry nor a center of symmetry.

Enantiomorphous crystals exhibit optical activity and are capable of rotating the plane of polarization of plane-polarized light. One form rotates to the left (levorotatory, or L-form), and one to the right (dextrorotatory, or D-form).

Molecules and substances that exhibit optical activity are described as chiral. Pasteur first isolated optical isomers of tartaric acid as shown in Figure 2.3. Mixtures of D and L crystals in solution will be optically inactive but can be resolved, unlike meso-form.

Crystalline racemates are normally considered to belong to one of two basic classes:

- Conglomerate: an equimolar mechanical mixture of two pure enantiomorphs;
- Racemic compound: an equimolar mixture of two enantiomers homogeneously distributed throughout the crystal lattice.

Racemates can be resolved manually or by forming a salt or an ester with an optically active base (usually an alkaloid) or alcohol followed by fractional distillation.

CRYSTAL HABIT

Most manufactured and natural crystals are distorted to some degree, and this may lead to misunderstanding of the term "symmetry." If crystal habit is considered as the product of restrictions on the growth of crystals in particular directions, then some generalizations can be made (8, 24). No restrictions result in a crystal simply being a larger version of the unit cell lattice structure. For example, a cubic lattice

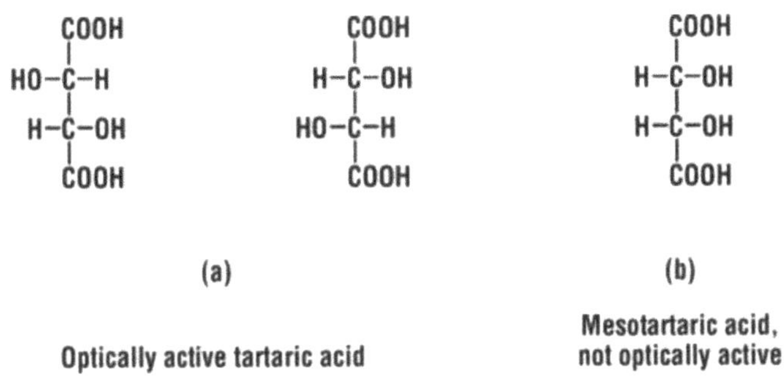

FIGURE 2.3 Isomers of tartaric acid.

will give rise to a large cubic crystal. Restrictions in one dimension will result in plate-like, tabular, or flaky particles. Restrictions in two dimensions yield needle-shaped acicular particles.

Rapid crystallization from supercooled melts, supersaturated solutions, and vapors frequently produce tree-like formations called dendrites. Most metals crystallize in this manner, but due to the filling-in process, the final crystalline mass may show little outward appearance of dendrite formation.

CRYSTAL IMPERFECTIONS

Most crystals are imperfect. Lattice imperfections and other defects can confer some important chemical and mechanical properties on crystalline materials. The three main types of lattice imperfection are point (zero-dimensional), line (one-dimensional), and surface (two-dimensional).

Vacancies are lattice sites from which units are missing, leaving "holes" in the structure. These units may be atoms (e.g., metallic, molecular, or ionic). Interstitial defects arise when foreign atoms occupy positions in the interstices between matrix atoms. Interstitial defects lead to lattice distortion. Complex point defects occur in ionic crystals. A cation can leave its site and become relocated interstitially near a neighboring cation. This combination (cation vacancy and interstitial cation) is called a Frenkel imperfection. A Schottky imperfection is a cation and anion vacancy. A substitutional impurity is a foreign atom that occupies the site of a matrix atom.

Two main types of line defects play a role in crystal growth: the edge and screw dislocations. Both of these are responsible for slip or shearing in crystals. Large numbers of dislocations occur in most crystals. Surface defects or imperfections involve the mismatch of boundaries and can be produced in crystalline materials as a result of mechanical or thermal stresses or irregular growth.

AMORPHOUS MATERIALS

Amorphous materials are those that have no underlying geometrical structure (25, 26). Unlike the crystalline materials described above, amorphous materials do not take on a symmetry defined by atomic or molecular interactions. They are epitomized by an unstructured random orientation of atoms or molecules (27). This randomness gives rise to two important properties that impact the behavior of the powder, its component particles, and the dosage form of which they are the foundation. Firstly, amorphous materials can restructure under the influence of pressure, temperature, or moisture changes to form crystalline materials or hydrates (28, 29). Consequently, they may be physically unstable (27, 30, 31). Also, their Gibbs free energy for dissolution is high rendering this state easier to dissolve than a crystalline form and at the same time more prone to attract water from the atmosphere, particularly at high humidity (20, 32, 33).

It is worth noting that the presence of amorphous material as a proportion of the total mass of a particle or powder may not be the most important metric of the likelihood of instability. The presence of amorphous material at the surface of particles is more likely to be an indicator of behavior (34). A high proportion of amorphous material found predominantly at the surface of particles might not translate to a high proportion of the total mass since the surface area to volume ratio is small. However, where the entire particle is intended to be amorphous, additional formulation steps, such as the addition of excipients, may be required to stabilize the particle structure (28, 35). Indeed, crystallinity within amorphous materials may contribute to the complexity of controlling it as a critical quality attribute (36).

Historically, the amorphous content in solid oral dosage forms was considered potentially detrimental to the performance and stability. However other dosage forms, such as inhaled aerosols, have employed entirely amorphous materials with environmental controls to prevent instability in storage and transport (34).

PARTICULATE SYSTEMS

Particulate systems are dispersed systems containing a solid phase with gas, liquid, or another solid (Table 2.3). Aerosols are solid particles or liquid droplets suspended in a gas. A complex aerosol may exist when solid particles are contained within the liquid droplets. A powder is a collection of small discrete solid particles in close contact with each other, with the void space usually filled with gas (37). Solid particles dispersed in a liquid phase are known as colloidal or coarse suspensions, according to the size of the dispersed particles. An emulsion contains liquid droplets dispersed in an immiscible liquid phase.

To understand the behavior of powders, the powder system needs to be conceived as a two-phase system of solid particles and gas. The interparticle forces and properties of the gas phase play important roles in powder dynamics (Rietema 1991).

Colloidal systems (also known as sols) contain particle sizes of 1 nm–0.5 μm (38, 39). The three types of colloidal systems, classified according to the interaction of dispersed particles with the dispersion medium, are lyophilic, lyophobic, and association colloids. Lyophilic ("solvent-loving") colloids disperse easily with the dispersion medium. The attraction between the dispersed and dispersion phases leads to solvation or hydration in aqueous media. Lyophobic ("solvent-hating") colloids have little attraction to the dispersion medium. These colloidal systems are produced by using high-energy dispersion or condensation methods. Association colloids contain surfactants arranged as micelles. Spherical micelles are produced at surfactant

TABLE 2.3
Types of Disperse Particulate Systems

	Continuous Phase		
Disperse Phase	**Gas**	**Liquid**	**Solid**
Gas	–	Foam	Solid foam
Liquid	Liquid aerosol	Emulsion	Solid emulsion
Solid	Solid aerosol	Suspension	Solid suspension

concentrations close to the critical micelle concentration (CMC). Laminar micelles occur at higher concentrations. Although micelles are not solid particles, their size allows them to be considered as colloidal dispersions. Both lyophilic and association colloidal systems are thermodynamically stable, whereas lyophobic colloidal systems are thermodynamically unstable.

The presence and magnitude of electrical charge on colloidal particles affect the stability of colloidal systems. Like charges on colloid particles produce repulsion and prevent agglomeration, thereby increasing the stability of the colloidal system. Oppositely charged colloid particles may aggregate when mixed.

Coarse suspensions contain particles larger than 0.5 μm (40, 41). Pharmaceutical suspensions are historically coarse dispersions, in which insoluble particles are dispersed in a liquid medium. The desirable properties of an acceptable suspension include a slow settling rate of dispersed particles, ease of redispersion into a uniform mixture by shaking following settling, and sufficient viscosity for its application. These qualities are achieved by flocculated particles. Flocculated particles are weakly bonded, settle rapidly, do not form a cake, and are easily resuspended. Deflocculated particles settle slowly, but they form a hard cake that is difficult to resuspend (40, 41).

CONCLUSION

The structure of crystalline materials has been characterized and can be defined in terms of 230 space lattices that fall into 32 classes of symmetry and 7 systems that, in addition to the amorphous state, describe the range of particle morphology. Since regular geometries are a major simplification of true particle morphologies, the effects of specific defects that modify crystal structures to give rise to irregular forms are acknowledged.

Particles are produced by a variety of methods. Crystallization occurs from saturated solutions or hot melts to form solids. A variety of drying methods may be used to evaporate solvent and induce supersaturation. Crystals can be prepared by homogeneous or heterogeneous nucleation from solution. In the latter case, seed nuclei may be employed to control the form of the final product. Particles can also be modified by milling to reduce particle size by using high-energy input processes. Once particles have been prepared, they can be characterized by surface analytical methods to classify their structures and define them prior to their inclusion in pharmaceutical products.

The following chapter presents an overview of the major methods of manufacture from which particles of specific sizes and with specific properties emerge. Subsequent chapters address these properties and their contribution to dosage form design and drug delivery. Particle morphology is a fundamental element in suspension and powder behavior. An understanding of the principles of particle formation and manufacturing process controls is essential for the optimization of product performance.

REFERENCES

1. Rukeyser M. *Willard Gibbs*. Woodbridge, CT: Oxbow Press; 1942.
2. Gibbs W. *The Scientific Papers of J. Willard Gibbs*. Woodbridge, CT: Oxbow Press; 1993.
3. Mullin J. *Crystallization*. Oxford, UK: Butterworth-Heinemann; 1993.
4. Glusker J, Lewis M, Rossi M. *Crystal Structure Analysis for Chemists and Biologists*. New York: VCH Publishers; 1994.
5. Hadjittofis E, Isbell MA, Karde V, Varghese S, Ghoroi C, Heng JYY. Influences of crystal anisotropy in pharmaceutical process development. *Pharm Res*. 2018;35:100. doi:10.1007/s11095-018-2374-9.
6. Jones W, March N. *Theoretical Solid State Physics 1. Perfect Lattices in Equilibrium*. Minneola, NY: Dover; 1985.
7. Jones W, March N. *Theoretical Solid State Physics 2. Non-equilibrium and Disorder*. Minneola, NY: Dover; 1985.
8. Tiwary AK. Modification of crystal habit and its role in dosage form performance. *Drug Dev Ind Pharm*. 2001;27:699–709.
9. Umprayn K, Luengtummuen A, Kitiyadisai C, Pornpiputsakul T. Modification of crystal habit of ibuprofen using the phase partition technique: Effect of Aerosil and Tween 80 in binding solvent. *Drug Dev Ind Pharm*. 2001;27:1047–56.
10. Bennema P. Growth forms of crystals: Possible implications for powder technology. *KONA Powder and Particle*. 1992;10:25–40.
11. Garekani H, Sadeghi F, Badiee A, Mostafa S, Rajabi-Siahboomi AR, Rajabi-Siahboomi A. Crystal habit modifications of ibuprofen and their physico-chemical characteristics. *Drug Dev Ind Pharm*. 2001;27:803–9.
12. Nuffield E. *X-ray Diffraction Methods*. New York, NY: Wiley; 1966.
13. Llacer J, Galardo V, Delgado R, Parraga J, Martin D, Ruiz M. X-ray diffraction and electron microscopy in the polymorphism study of Ondasetron hydrochloride. *Drug Dev Ind Pharm*. 2001;27:899–908.
14. Darcy P, Buckton G. Crystallization of bulk samples of partially amorphous spray-dried lactose. *Pharm Dev Tech*. 1998;3:503–7.
15. Brittain H. Pharmaceutical applications of X-ray powder diffraction. *Pharm Tech*. 2001;25:142–50.
16. Be HG. *Polymorphism in Pharmaceutical Solids*. New York, NY: Informa Healthcare; 2009.
17. Turi E, Khanna Y, Taylor T. Thermal analysis. In: Sibilia J, editor. *A Guide to Materials Characterization and Chemical Analysis*. New York, NY: VCH Publishers; 1988. p. 205–28.
18. Okonogi S, Puttipipatkhachorn S, Yamamoto K. Thermal behavior of ursodeoxycholic acid in urea: Identification of anomalous peak in the thermal analysis. *Drug Dev Ind Pharm*. 2001;27:819–23.
19. Mackenzie R. *Differential Thermal Analysis*. London, UK: Academic Press; 1970.
20. Hsieh DD, Roberts D, Rosner T, Rosenbaum T, Lai C, Gao Q. Combining solid-state and solution calorimetry for the Gibbs free energy calculation of polymorphs to determine the relative stability. *Crystal Growth Des*. 2016;16:12–24. doi:10.1021/acs.cgd.5b00443.
21. Seyer J, Luner P. Determination of indomethacin crystallinity in the presence of excipients using diffuse reflectance near-infrared spectroscopy. *DPharm Dev Tech*. 2001;6:573–82.
22. Luner P, Majuru S, Seyer J, Kemper M. Quantifying crystalline form composition in binary powder mixtures using near-infrared reflectance spectroscopy. *Pharm Dev Tech*. 2000;111:231–43.

23. Buckton G, Darcy P. Assessment of disorder in crystalline powders—a review of analytical techniques and their application. *Int J Pharm.* 1999;179:141–58.
24. Braga D, Casali L, Grepioni F. The relevance of crystal forms in the pharmaceutical field: Sword of Damocles or innovation tool. *Int J Mol Sci.* 2022;23:9013.
25. Yu L. Amorphous pharmaceutical solids: Preparation, characterization and stabilization. *Adv Drug Deliv Rev.* 2001;48:27–42.
26. Bauer JF. Pharmaceutical solids—The amorphous phase. *J Valdi Technol.* 2009;19:63–8.
27. Wang Y, Wong Y, Cheng J, Chen H, Xu J, Liu J, et al. Recent advances in the application of characterization techniques for studying physical stability of amorphous solids. *Crystals.* 2021;11:1440. doi: 10.3390/cryst11121440.
28. Laitinen R, Lobmann K, Strachan CJ, Grohganz H. Emerging trends in the stabilization of amorphous drugs. *Int J Pharm.* 2012;453:65–79.
29. Jelic D. Thermal stability of amorphous solid dispersions. *Molecules.* 2021;26:238. doi: 10.3390/molecules26010238.
30. Edueng K, Mahlin D, Bergstrom CAS. The need for restructuring the disordered science of amorphous drug formulations. *Pharm Res.* 2017;34:1754–72.
31. Alhalaweh A, Alzghoul A, Mahlin D, Bergstrom CAS. Physical stability of drugs after storage above and below the glass transition temperature: Relationship to glass forming ability. *Int J Pharm.* 2015;495:312–7. doi:10.1016/j.ijpharm.2015.08.101.
32. Skymer RE, McDonagh JL, Groom CR, Mourik Tv, Mitchell JBO. A review of methods for the calculation of solution free energies and the modelling of systems in solution. *PhysChem Chem Phys.* 2015;17:6174. doi: 10.1039/c5cp00288e.
33. Lattye A. Crystal dissolution kinetics and Gibbs free energy. *J Electron Spectroscopy and Related Phenomena.* 2006;180:248–59.
34. Hickey AJ, Mansour HM, Telko MJ, Zhou Z, Smyth HD, Mulder T, et al. Physical characterization of component particles included in dry powder inhalers. I. Strategy review and static characteristics. *J Pharm Sci.* 2007;96:1282–301. doi: 10.1002/jps.20916.
35. Amstad E, Spaepen FA, Weitz DA. Stabilization of the amorphous structure of spray-dried drug nanoparticles. *J Phys Chem B.* 2016;120:9161–5.
36. Moseson DE, Taylor LS. Crystallinity: A complex critical quality attribute of amorphous solid dispersions. *Mol Pharmaceut.* 2023;20:4802–25.
37. Rietema K. *The Dynamics of Fine Particles.* New York, NY: Elsevier; 1991.
38. Yee BM, Sinko PJ. Colloidal dispersions and nanometrics. In: Sinko PJ, editor. *Martin's Physical Pharmacy and Pharmaceutical Sciences.* Philadelphia, PA: Wolters Kluwer; 2023. p. 577–608.
39. Garg T, Rath G, Goyal AK. Colloidal drug delivery systems: Current status and future directions. *Crit Rev Ther Drug Carrier Syst.* 2015;32:89–147.
40. Hiestand EN. Physical properties of coarse suspensions. *J Pharm Sci.* 1972;61:268–72.
41. Yee BM, Sinko PJ. Coarse dispersions. In: Sinko PJ, editor. *Martin's Physical Pharmacy and Pharmaceutical Sciences.* Philadelphia, PA: Wolters Kluwer; 2023. p. 551–76.

3 Primary Particle Manufacturing

PARTICLE PRODUCTION

Pharmaceutical particulates may be produced by constructive and destructive methods, as depicted in Figure 3.1 (1, 2). Each of these methods results in a particulate system characterized by defined particle size distribution, surface area, and surface properties. Particle constructive methods include crystallization (3–6), freeze-drying (7–12), and supercritical fluid techniques (13, 14). Particle destructive methods include milling and grinding (15). Spray drying may be considered a unique combination of destructive and constructive methods since the spraying involves the dispersion of droplets from a solution or suspension, thereby creating a surface in a destructive manner while allowing the assembly of particles from molecular or particulate materials contained in each droplet in a constructive manner (6, 13, 16–20).

Figure 3.1 places the major methods of manufacture in the context of their origins, either by constructing particles from the molecules, constructive, or by reducing starting solid or liquid state materials to desired sizes, destructive. The key properties that are controlled are the final particle size and distribution and the related property of surface area, and this is brought about through energy expenditure which for destructive methods requires mechanical energy input and for constructive methods involves energy imparted through thermal, pressure, or chemical modulation. Since destructive methods involve the creation of new surface that is not controlled by molecular mechanisms, they give rise to more variable surface chemistry and properties that require monitoring and potentially further processing control to support reproducible behavior in the dosage form.

Often a manufacturing method is a combination of a destructive and a constructive method. For example, spraying and mixing are methods used for creating dispersed phases (droplets) within a surrounding medium (gas or liquid); within these dispersed units, constructive crystallization, precipitation, or polymerization may be achieved by the removal of the solvent from the dispersed phase.

METHODS—CONSTRUCTIVE

Crystallization is the most common method of particle production (21, 22). The solid crystalline state is produced from a liquid solution by cooling, evaporation, precipitation, or the addition of another compound (solvent or solute) (23). The process of crystallization involves nucleation followed by crystal growth. The crystalline

DOI: 10.1201/9781003470151-3

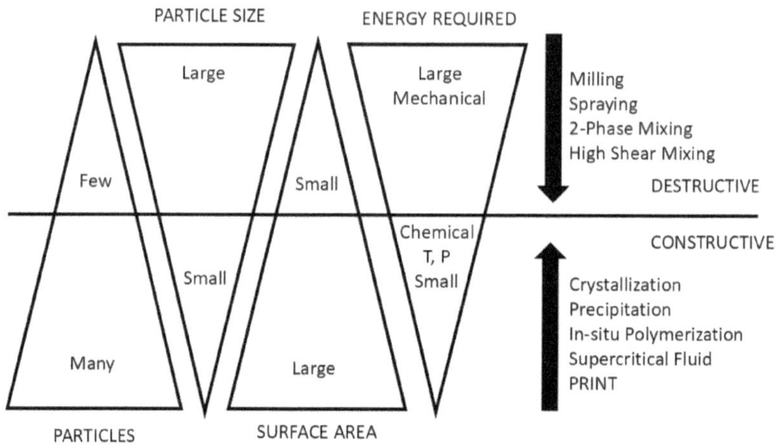

FIGURE 3.1 Destructive and constructive methods of particle manufacture and their physicochemical origins. The intent of producing particles of the desired size and distribution is achieved through mechanical, thermal (T), pressure (P), or chemical energy expenditure.

structure of particles is affected by solvent, solute concentration, cooling rate, stirring, and the presence of additives.

Crystallization from solution essentially involves reaching a critical concentration in the solution, which represents significant supersaturation for spontaneous nucleation to occur in a homogeneous solution. In the presence of foreign particles (primary heterogenous nucleation) or nucleating crystals (secondary nucleation), marginal supersaturation is required. Figure 3.2 charts these nucleation situations.

Freeze-drying, or lyophilization, is a process that involves freezing, sublimation, and secondary drying (24–27). Sublimation occurs when the water vapor partial pressure remains below the vapor pressure of ice. The solution composition, freezing temperature, and rate of freezing affect the sites and number of ice spaces. Sublimation is generally achieved by increasing the temperature at low pressure. The sublimation temperature should be optimized for the rate of sublimation, without the risk of melting. The sublimation pressure is adjusted to promote convection,

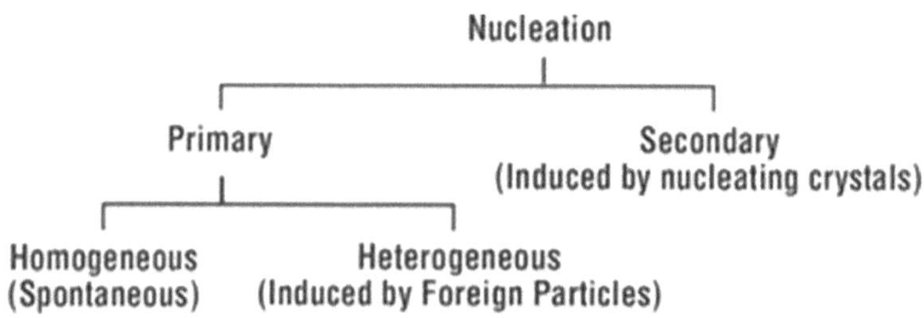

FIGURE 3.2 Crystallization from solution.

generally 0.3 times the vapor pressure of ice at the temperature used (28). The pressure is reduced in the secondary drying phase to eliminate adsorbed water.

Particle generation using supercritical fluids provides recrystallization by precipitation (13, 29–31). Supercritical fluids exist at a region of high temperature and pressure at which the gas and liquid phase have the same density and appear as a single phase. A commonly used supercritical fluid is carbon dioxide, which has a low critical temperature (31°C) and is nontoxic, nonflammable, and inexpensive. Particle generation by supercritical fluid includes rapid expansion of supercritical solutions (RESS) and supercritical antisolvent (SAS) techniques.

PRINT technology is a unique combination of spray drying and microextrusion where molds are employed to define the size and shape of particles, and following the drying process, they are collected for reuse in the specific dosage form application (32, 33).

METHODS—DESTRUCTIVE

Particle size reduction is generally performed by milling (1, 34, 35). The rate of size reduction depends on the material properties of the particle, starting size, the orientation in the mill, and the mill residence time. Compression, shear, and tension forces are used for size reduction (36). Compression forces produce crushing, shear forces produce cutting, and tension forces cause elongation and pull apart particles. Milled particles contain flaws, discontinuities, or imperfections in the structure. The application of milling forces at the flaws causes cracks and cleavage of the particles. Various types of milling equipment exist for size reduction of particles (36), including the fluid energy mill, ball mill, and hammer mill. The final particle size depends upon the energy input of the milling process. Fluid energy mills produce 5–30 μm particles, ball mills produce 10–200 μm particles, and hammer mills produce 50–400 μm particles. Particles prepared by milling mostly originate in crystalline active pharmaceutical products and result in smaller crystalline particles. However, the mechanical disruption of the crystal structure can result in the presence of amorphous material that must be controlled or eliminated to assure stability in subsequent processing, storage, and transport (37).

Spray drying is a process by which a dried particulate form is produced by spraying a liquid into a hot drying medium (13, 38–40). It involves atomization of the liquid feed into fine droplets, evaporation of the solvent, and size separation of dried particles from the drying medium. Factors affecting the particle size, shape, and density of the final product include the feed concentration, solvent type, feed rate, atomization method, drying temperature, and residence time in the drying chamber. The process involves spraying a solution or suspension through a nozzle at a controlled liquid feed rate with nozzle airflow that disperses the liquid into droplets into a drying air stream. This is a continuous process that allows process control on liquid feed rate, nozzle and drying air rate, and inlet air temperature with monitoring of outlet air temperature. After traversing a drying tower, the dried particles pass through a cyclone separator operating as an elutriator, and particles are collected by impaction according to size. The airstream then passes ultrafine particles to a filter and vents the carrier gas. Where organic solvents are used, a condensation recovery system is arranged at the outlet. The input parameters can be controlled to achieve the desired particle size characteristics. Spray-dried particles can be prepared with a range of morphology and density based on input parameters. These may be selected for their desirability to the final application. The final product is often amorphous. Consequently, it may require the presence of excipients or packaging considerations (use of desiccant or barrier wraps) to ensure the stability of the product on storage and transport.

Materials that cannot tolerate the high temperatures employed in conventional spray drying are candidates for spray freeze-drying (32, 41, 42). Unlike spray drying, spray freeze-drying involves spraying droplets into a cryogenic agent such as liquid nitrogen and then recovering the droplets for secondary drying by lyophilization.

Thin film freezing is similar to spray drying in that it is a combination of constructive and destructive methods but operates with much lower energy input (43, 44). Large drops of solution or suspension of drugs and additives are deposited on a rotating drum operating at cryogenic temperatures. The large particles formed are collected after the sublimation of a portion of the solvent. The sub-structure within the particles is fiber-like with significant void-space. Mild milling breaks these large particles into micron-sized structures that are then dried in a lyophilizer. Particles can be generated with crystalline or amorphous structures, depending on the conditions used during their manufacture. Therefore, considerations may vary according to the final product structure as described for other methods above.

Counter-current flash precipitation is a variation on crystallization in that by passing a solution of drug and additives into an opposing flow of a non-solvent, particles are formed in discrete sizes controlled by flow rates, solvent properties, temperature, and pressure (34, 45).

COMPLEX MANUFACTURING PROCESSES

It would be hard to do justice to the enormous amount of research and development that has been conducted on complex methods of production of particles, particularly those using polymer as a means of controlling the delivery of drug (46). However, they can be listed as in-situ polymerization, double emulsion precipitation (47–49), and nano milling (50, 51). Additionally, particle surfaces can be modified by coacervation, forced control agents, and other methods of coating (26, 52–54). Often particles manufactured by these methods have an extra dimension of control based on the extended release of the drug. If this is the objective, the method of manufacture adopted has a dominant constructive element that gives the desired internal structure to support exquisitely controlled drug release. These particle engineering technologies have significantly increased the number of tools available today for drug developers to achieve a desired target product profile.

CONCLUSION

Particles are produced by a variety of methods. Crystallization occurs from saturated solutions or hot melts to form solids. A variety of drying methods may be used to evaporate solvent and induce supersaturation. Crystals can be prepared by homogeneous or heterogeneous nucleation from solution. In the latter case, seed nuclei may be employed to control the form of the final product. Particles can also be modified by milling to reduce their particle size using high-energy input processes. Once particles have been prepared, they can be characterized by surface analytical methods to classify their structures and define them prior to their inclusion in pharmaceutical products.

The following chapters present the appropriate methods of sampling to allow for representative characterization (Chapter 4) followed by critical particle population characteristics and their measurement as a means of predicting assuring performance. These criteria can then be linked to the quality, efficacy, and safety of the product.

Particle morphology is a fundamental element in suspension and powder behavior. An understanding of the principles of particle formation and manufacturing process controls is essential for the optimization of product performance.

REFERENCES

1. Hickey AJ. Lung deposition and clearance: What can be learned from inhalation toxicology and industrial hygiene? *Aerosol Science and Technology.* 1993;18:290–304.
2. Hickey AJ. *Pharmaceutical Process Engineering.* 3rd ed. Boca Raton, FL: CRC Press; 2024.
3. Tiwary AK. Modification of crystal habit and its role in dosage form performance. *Drug Dev Ind Pharm.* 2001;27:699–709.
4. Umprayn K, Luengtummuen A, Kitiyadisai C, Pornpiputsakul T. Modification of crystal habit of ibuprofen using the phase partition technique: Effect of Aerosil and Tween 80 in binding solvent. *Drug Dev Ind Pharm.* 2001;27:1047–56.
5. Tadayyan A, Rohani S. Modeling and nonlinear control of continuous KCl-NaCl crystallizer. *Part Sci Tech.* 2000;18:329–57.
6. Pham H, Luo P, Genin F, Dash A. Synthesis and characterization of hydroxyapatite-ciprofloxacin delivery systems by precipitation and spray drying technique. *AAPS PharSciTech.* 2001;3:Article 1.
7. Vavia P, Adhage N. Freeze-dried inclusion complexes of tolfenamic acid and beta cyclodextrins. *Pharm Dev Tech.* 2000;5:571–4.
8. Rey L, May J. *Freeze-drying/lyophilization of pharmaceutical and biological products.* New York, NY: Marcel Dekker; 1999.
9. Nuijen B, Nuijen B, Bouma M, Talsma H, Manada C, Jimeno J, et al. Development of a lyophilized parenteral pharmaceutical formulation of the investigational polypeptide marine anticancer agent Kahak=lalide F. *Drug Dev Ind Pharm.* 2001;27:767–80.
10. Kirsch L, Zhang S, Muangsiri W, Redmon M, Luner P, Wurster D. Development of a lyophilized formulation for (R,R)-formoterol (L)-tartrate. *Drug Dev Ind Pharm.* 2001;27:89–96.
11. Jameel F, Amberry K, Pikal M. Freeze drying properties of some oligonucleotides. *Pharm Dev Tech.* 2001;6:151–7.

12. Jaeghere F, Allemann E, Feijen J, Kissel T, Doelker E, Gurny R. Freeze-drying and lyopreservation of diblock and triblock poly(lactic acid)-poly(ethylene oxide) (PLA-PEO) copolymer nanoparticles. *Pharm Dev Tech.* 2000;5:473–83.
13. Sacchetti M, van Oort MM. Spray drying and supercritical fluid particle generation techniques. In: Hickey AJ, editor. *Inhalation Aerosols: Physical and Biological Basis for Therapy.* New York, NY: Marcel Dekker; 1996. p. 337–85.
14. Rogers T, Johnston K, Williams R. Solution-based particle formation of pharmaceutical powders by supercritical or compressed fluid CO_2 and cryogenic spray freezing technologies. *Drug Dev Ind Pharm.* 1998;27:1003–15.
15. Mura P, Faucci M, Parrini P. Effects of grinding with microcrystalline cellulose and cyclodextrins on the ketoprofen physico-chemical properties. *Drug Dev Ind Pharm.* 2001;27:119–28.
16. Takeuchi H, Yasuji T, Yamamoto H, Kawashima Y. Temperature and moisture-induced crystallization of amorphous lactose in composite particles with sodium alginate prepared by spray drying. *Pharm Dev Tech.* 2000;5:355–63.
17. Kumar V, Kang J, Yang T. Preparation and characterization of spray-dried oxidized cellulose microparticles. *Pharm Dev Tech.* 2001;6:449–58.
18. Esposito E, Roncarati R, Cortesi R, Cervellati F, Nastruzzi C. Production of Eudragit microparticles by spray drying technique: Influence of experimental parameters on morphological and dimensional characteristics. *Pharm Dev Tech.* 2000;5:267–78.
19. Dunbar C, Concessio N, Hickey AJ. Evaluation of atomizer performance in production of respirable spray-dried particles. *Pharm Dev Tech.* 1998;31:433–41.
20. Carrigy N, Vehring R. Engineering stable spray-dried biologic powder for inhalation. In: Hickey AJ, Rocha SRd, editors. *Pharmaceutical Inhalation Aerosol Technology.* 3rd ed. Boca Raton, FL: CRC Press; 2019. p. 291–326.
21. Orehek J, Teslia D, Likozar B. Continuous crystallization processes in pharmaceutical manufacturing: A review. *Organic Process Research and Development.* 2021;25:16–42.
22. Gao Z, Rohani S, Gong J, Wang J. Recent developments in the crystallization process: Toward the pharmaceutical industry. *Engineering.* 2017;3:343–53.
23. David R, Giron D. Crystallization. In: Chulia D, Deleuil M, Pourcelot Y, editors. *Powder Technology and Pharmaceutical Processes.* Amsterdam, The Netherlands: Elsevier; 1994.
24. Patel SM, Pikal MJ. Emerging freeze-drying process development and scale-up issues. *AAPS PharSciTech.* 2011;12:372–8.
25. Nail SJ, Jiang S, Chongprasert S, Knopp SA. Fundamentals of freeze drying. *Pharm Biotech.* 2002;14:281–360.
26. Timilsena YP, Akanbi TO, Khalid N, Adikar B, Barrow CJ. Complex coacervation: Principles, mechanisms and applications in microencapsulation. *Int J Biological Macromolecules.* 2019;121:1276–86.
27. Tchessalov S, Maglio V, Kazarin P, Alexeenko A, Bhatnagar B, Shani E, et al. Practical advice on scientific design of freeze-drying process: 2023 update. *Pharm Res.* 2023;40:2433–55.
28. Bentejac R. Freeze drying. In: Chulia D, Deleuil M, Pourcelot Y, editors. *Powder Technology and Pharmaceutical Processes.* Amsterdam, The Netherlands: Elsevier; 1994. p. 257–71.
29. Girotra P, Singh SK, Nagpal K. Supercritical fluid: A promising approach in pharmaceutical research. *Pharm Dev Tech.* 2013;18:22–58.
30. Park H, Kin J-S, Kim S, Hwang SJ. Pharmaceutical applications of supercritical fluid extraction of emulsions for micro-/nanoparticle formulation. *Pharmaceutics.* 2021;13:1928.

31. Aguiar-Ricardo A, Costa E. Supercritical fluid manufacture. In: Hickey AJ, Rocha SRd, editors. *Pharmaceutical Inhalation Aerosol Technology*. 3rd ed. Boca Raton, FL: CRC Press; 2019. p. 327–48.
32. Xu J, Wong DHC, Byrne JD, Chen K, Bowerman C, DeSimone JM. Future of particle replication in nonwetting templates (PRINT) technology. *Angew Chem Int Ed Engl*. 2013;52:6580–9.
33. Wanning S, Suverkrup R, Lamprecht A. Pharmaceutical spray freeze drying. *Int J Pharm*. 2015;488:136–53.
34. Tao J, Chow SF, Zheng Y. Application of flash nanoprecipitation to fabricate poorly water-soluble drug nanoparticles. *Acta Pharmaceutica Sinica B*. 2019;9:4–18.
35. Tan BMJ, Chan LW, Heng PWS. Milling and blending: Producing the right particles and blend characteristics for dry powder inhalation. In: Hickey AJ, Rocha SRd, editors. *Pharmaceutical Inhalation Aerosol Technology*. 3rd ed. Boca Raton, FL: CRC Press; 2019. p. 273–90.
36. Lantz R. Size reduction. In: Lieberman H, Lachman L, editors. *Pharmaceutical Dosage Forms*. New York, NY: Marcel Dekker; 1981. p. 1–53.
37. Challener CA. Stabilization of amorphous API. *Pharm Tech*. 2021;45:20–4.
38. Baumann JM, Adam MS, Wood JD. Engineering advances in spray drying for pharmaceuticals. *Ann Rev Chemical and Biomolecular Engineering*. 2021;12:217–40.
39. Pinto JT, Faulhammer E, Dieplinger J, Dekner M, Makert C, Nieder M, et al. Progress in spray-drying of protein pharmaceuticals: Literature analysis of trends in formulation and process attributes. *Drying Technology*. 2021;39:1415–46.
40. Vehring R. Pharmaceutical particle engineering via spray drying. *Pharm Res*. 2007;25:999–1022.
41. Farinha S, Sa JV, Lino PR, Galesio M, Pires J, Rodrigues MA, et al. Spray freeze drying of biologics: A review and applications for inhalation delivery. *Pharm Res*. 2022;40:1115–40.
42. Aolali MB, Barresi AA, Boccardo G, Pisano R. Spray-freeze-drying as a solution to continuous manufacturing of pharmaceutical products in bulk. *Processes*. 2020;8:709.
43. Overhoff KA, Johnston KR, Tam J, Engstrom J, Williams III RO. Use of thin film freezing to enable drug delivery: A review. *J Drug Deliv Sci Tech*. 2009;19:89–98.
44. Hufnagel S, Sahakijpijarn S, Moon C, Cui Z, Williams III RO. The development of thin-film freezing and its application to improve delivery of biologics as dry powder aerosols. *KONA Powder and Particle*. 2022;39:176–92.
45. Saad WS, Prud'homme RK. Principles of nanoparticle formation by flash nanoprecipitation. *Nanotoday*. 2016;11:212–27.
46. Burgess DJ, Hickey AJ. Microsphere technology and applications. In: Boylan J, Swarbrick J, editors. *Encyclopedia of Pharmaceutical Technology*. New York, NY: Marcel Dekker; 1996. p. 441–76.
47. Vilos C, Velasquez LA. Therapeutic strategies based on polymeric microparticles. *J Biomed Biotechol*. 2012;2012:672760.
48. Bhujel R, Maharjan R, Kim NA, Jeong SH. Practical quality attributes of polymeric microparticles with current understanding and future perspectives. *J Drug Deliv Sci Tech*. 2021;64:102608.
49. Lagreca E, Onesto V, DiNatale C, LaManna S, Netti PA, Vecchione R. Recent advances in the formulation of PLGA microparticles for controlled drug delivery. *Progress in Biomaterials*. 2020;9:153–74.
50. Li M, Azad M, Dave R, Bilgili E. Nanomilling of drugs for bioavailability enhancement: A holistic formulation-process perspective. *Pharmaceutics*. 2016;8:17.

51. Peltonen L, Hirvonen J. Pharmaceutical nanocrystals by nanomilling: Critical process parameters, particle fracturing and stabilization methods. *J Pharm Pharmacol.* 2010;62:1569–79.
52. Sing CE, Perry SL. Recent progress in the science of complex coacervation. *Soft Matter.* 2020;16:2885–914.
53. Ban E, Kim A. Coacervation: Recent developments as nanostructure delivery platforms for therapeutic biomolecules. *Int J Pharm.* 2022;624:122058.
54. Bungert N, Kobler M, Sherleiss R. Proof-of-concept for adjusted surface energies and modified fines as a novel concept in particle engineering for DPI formulations. *Pharmaceutics.* 2022;14:14050951.

4 Sampling Techniques, Errors, and Measurement Bias

INTRODUCTION

Sampling is the process of selecting a small part of a population to estimate a given characteristic of the entire population. Each population contains a finite number of identifiable units, each with a fixed value for a particular characteristic. The population characteristics can be described by statistical terms including the population mean (μ), standard deviation (σ), and variance (σ^2). The population characteristics would be known if the entire population was sampled and accurately measured. Sampling is used to estimate the population parameters. The sample statistics include the sample mean ($\bar{x}$), standard deviation (s), variance (s^2), standard error of the mean (SEM), and coefficient of variance (CV). The statistics obtained from the sample are variable and depend on each selected sample as well as the variability of the measurement (1, 2). Each sample will provide an estimate of the population's characteristics, but sample results will differ slightly from each other. The Central Limit Theorem states that sample means follow a normal (Gaussian) distribution regardless of the shape of the original distribution if the sample size is large enough. This allows inferences to be made about a population based on the sample mean.

Representative sampling plays a critical role in ensuring the quality, safety, and efficacy of pharmaceutical products across the product lifecycle. During product development, it is important to ensure that representative samples are captured so that accurate decisions are made for the formulation and manufacturing process. Representative sampling allows batch-to-batch consistency to be assessed for raw materials, intermediates, and finished products. Representative samples are required according to Good Manufacturing Practices (GMPs) to ensure that the samples tested accurately reflect the characteristics of the entire batch. This is essential for quality control, quality assurance, and regulatory compliance of pharmaceutical products for clinical or commercial use.

A sample is considered representative when it accurately reflects the properties and characteristics of the larger population from which it was taken. The sample should possess the same characteristics as those of the population so that the measurements performed on it can be extrapolated to the entire population. Samples are taken for physical and chemical analysis of raw materials, intermediates, and finished products. For example, raw materials are sampled and tested for identification,

DOI: 10.1201/9781003470151-4

assay, potency, impurities, moisture content, and particle size distribution. In-process materials may be sampled and tested for in-process tests, such as content uniformity, moisture content, bulk density, and particle size distribution. Final products are sampled for process validation, product release and stability testing, and GMP retention samples. Batch disposition decisions are made based on the assumption that the sample taken accurately estimates the population. These decisions may occur during the research and development stage, where product and process data understanding is gathered, or during clinical or commercial manufacturing, where the batch may be released for patient use.

Based on Gy's definition, representative sampling occurs when the following two conditions are fulfilled: (1) probabilistic sampling occurs and (2) sample integrity is respected (3–7). Probabilistic sampling occurs when each unit has an equal probability of being selected, based on the principles of statistical random selection. Sample integrity means that the fundamental characteristics and properties of the sample are maintained during the sampling process. The sample was collected without biases, losses, alterations, or contamination. Representative sampling provides samples that accurately reflect the characteristics of the bulk population.

There may be cases where probabilistic sampling occurred, but the sample integrity was not maintained. For example, the particles may have been crushed during sampling, resulting in smaller particle sizes or differing shapes, compared with the bulk population. Care must be taken to ensure that the sampling process has not affected the intrinsic properties of the sample. There may also be cases where the sample integrity was maintained but the sampling was not fully probabilistic. For example, a suspension may have settled prior to sampling, resulting in a higher concentration at the bottom of a vessel. The sample collected from a dip tube located at the middle level of the vessel would not fully represent the suspension concentrations at the top or bottom layers.

In contrast, non-probabilistic sampling is based on the sampler's judgment or discretion. It generally arises from convenience or simplicity rather than principles of statistical probability. An example of non-probabilistic sampling is scooping a sample from the top of a large container. The samples obtained from non-probabilistic sampling may be biased and non-representative due to uncontrolled or unknown variables. Non-probabilistic sampling is not recommended but sometimes cannot be avoided.

Special care must be taken during the sampling of particulate systems. Unlike miscible liquids, which form homogeneous mixtures at the molecular level, all particulate materials exhibit some degree of heterogeneity. There are two types of heterogeneity: constitutional and distributional heterogeneity. Constitutional heterogeneity refers to the intrinsic variations observed between individual components. This refers to the variations in chemical or physical characteristics of individual particles, such as chemical composition, crystalline structure, particle size, or shape. Distributional heterogeneity refers to the variations in the spatial arrangement of particulate components within the bulk population. These variations may be due to segregation, aggregation, or sedimentation. Adequate mixing of the bulk material can minimize the effects of distributional heterogeneity.

For a generalized framework, particulate systems may be sampled from a stationary heap or a flowing stream (also known as process sampling). Many authors state that sampling from a stationary heap is not recommended and that powders should only be sampled while in motion with the whole stream taken for short periods of time rather than a portion of the stream for the entire time (8, 9). However, sampling from stationary heaps is common practice in the pharmaceutical field. This includes the sampling of APIs, raw materials, or intermediate products from bags, drying trays, blenders, and in-process storage vessels. There are some cases where samples are obtained from a flowing stream, such as sampling from a discharge chute. Examples include sampling of pellets from an extruder, granules from a roller compactor, tablets from a compression machine, or capsules from an encapsulation machine.

The main challenge of sampling from a stationary heap is segregation and its impact on the distributional heterogeneity of the particulate system. Segregation occurs when particles have freedom of movement and differences in physical properties, such as size or density. It is affected by particle size distribution, particle density, particle shape, coefficient of friction, surface roughness, and moisture content. The segregation potential of powders can be predicted by their flow properties. Free-flowing powders have the highest tendency to segregate, whereas cohesive powders have a lower tendency to segregate due to stronger inter-particulate forces. During segregation, particles with the same physical properties tend to accumulate together (10–13).

Segregation may occur as the heap is being formed and during transport or storage. As powder is poured into a heap, the larger particles may roll down the outer inclined surface, leaving the fine particles in the central region of the cone. As powders are poured into a container, the air is displaced. This causes finer particles to become suspended in the upward-moving air while the larger particles settle. This results in the generation of a plume of fine particles caused by elutriation segregation. Segregation may occur during powder handling, storage, or transport due to vibrations or gravity. Larger particles move towards the upper surface of the powder bed, while smaller particles fall through the gaps and are compacted at the bottom of the container. Greater segregation is observed at higher vibrational amplitudes and lower frequencies.

SAMPLING FROM STATIONARY HEAPS

Sampling techniques for obtaining particulate samples from stationary heaps include grab, composite, and incremental sampling. These samples are usually obtained with a scoop, spatula, or sampling thief.

Grab sampling (also known as spot sampling) involves collecting a single sample from the stationary pile. This is a non-probabilistic sampling method, and the sample is unlikely to be fully representative of the entire population because the sample is typically collected from a convenient and easy-to-access location, such as the top surface. Variations within the material characteristics (distributional heterogeneity) are not captured by a single grab sample. In populations with a high degree of distributional heterogeneity, grab sampling is likely to produce a larger mismatch between the individual sample result and the bulk population result (14). However, grab sampling may be adequate for specific types of testing. For example, a grab sampling of incoming raw materials is generally acceptable when identity is the only test performed.

Composite sampling involves collecting and combining multiple grab samples obtained from different locations of the stationary pile. Several grab samples are mixed to provide a single composite sample, generating a more representative sample and accounting for some degree of distributional heterogeneity. Composite sampling compensates for the heterogeneity in the bulk population. Smaller variability is observed with composite sampling compared with grab sampling for the same bulk population (14). An example of composite sampling is obtaining a powder blend sample for bulk assay, bulk density, particle size distribution, or moisture content testing.

Incremental sampling involves collecting multiple samples obtained at different locations within the stationary pile to assess the distributional heterogeneity. It provides a more comprehensive understanding of the population characteristics. Samples are obtained at different predetermined spatial locations. Incremental sampling is more time-consuming and labor-intensive than grab or composite sampling. An example of incremental sampling from a stationary pile is taking blend samples from a blender using a sample thief to assess the content uniformity.

Sample scoops and thieves are common tools used to collect samples from stationary heaps. Sample scoops collect samples from the surface of the powder bed, whereas thieves collect samples from the body of the powder bed. As noted earlier, sampling taken from the top surface of a powder bed is a non-probabilistic sampling technique and may not provide a sample that is fully representative of the bulk population. Sampling biases are common for both tools.

Sampling biases observed from using a scoop may result in differences in the observed characteristics of the sample. Examples of sampling biases include operator bias, where scooped samples may be selected due to the operator's selective judgement; sample location and/or depth, where the sample characteristics are affected by segregation (e.g., samples from upper levels may consist of larger particles); scoop size (e.g., small scoops may not be able to capture large particles); or contamination (e.g., contamination from materials located on the upper surface of the powder bed when trying to obtain samples at lower levels).

There are generally two types of sample thieves: side-sampling and end-sampling thieves (15–17) (Figure 4.1). Side-sampling thieves, also known as grain thieves, are used to sample free-flowing powders, whereas end-sampling thieves, also known as slug or plug thieves, are used to sample compressible powders. Side-sampling thieves may contain a single or multiple sample chambers, whereas end-sampling thieves only contain a single sample chamber. Side-sampling thieves have an inner rod containing one or more sample cavities and an outer rotating sleeve with openings aligned to the sample cavities. The outer tube is twisted to an open or closed position. Free-flowing particles flow into the opened cavities and are captured when

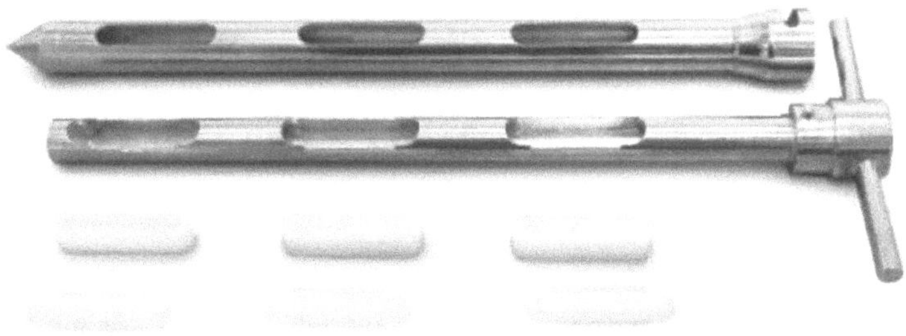

FIGURE 4.1 Multiport side-sampling thief (Accupod®) illustrating a probe (above) inserted into the powder and the internal chamber (below) to isolate the sample within the probe by turning to close the ports before withdrawing the sample.

the outer tube is twisted to the closed position. In contrast, the end-sampling thief has an inner rod and outer tube with a single sample cavity at the proximal tip. The inner rod is retracted to create a sample cavity at the tip. The cavity size may be minimally adjusted and then locked into a fixed position with a screw mechanism. The end-sampling thief is inserted further into the powder bed in a closed position. At the pre-determined location, the inner rod is retracted to expose the sample cavity. The thief is pushed deeper into the powder bed to capture the sample by powder compression, resulting in a sample plug. Since there is no sample closure mechanism, care must be taken not to dislodge the sample plug when the end-sampling thief is retracted from the powder bed. Free-flowing powders that cannot adequately compress to form a plug are unable to be sampled using the end-sampling thief. For both thief types, the sample size (weight) is dictated by the volume of the sample chamber. Thus, powders with higher density will result in a higher sample weight. Both sampling thief types generally have a tapered tip to minimize powder disruption when introduced into the powder bed.

Sampling biases from using a sampling thief are widely documented (13, 15, 16, 18–21). Thief insertion into the powder bed may result in powder movement, segregation, or particle attrition. Particulate contamination from the upper regions of the bed to the lower regions may occur. Larger particles may preferentially flow into the sampling cavity of side-sampling thieves. The angle of insertion and spatial orientation of the cavity opening may affect the powder flow into the sample cavity. Powder flow may be reduced at lower regions of the powder bed due to greater static pressure or powder compaction caused by thief insertion. Free-flowing particles may enter the space between the outer tube and the inner rod before the sample cavity is opened. Particles may become entrapped in the space adjacent to the closed sample chamber and become captured as part of the sample.

Novel sample thief designs have been developed to improve sampling and reduce bias (20, 22, 23). However, these novel sampling thieves are generally not widely used.

SAMPLING FROM A FLOWING STREAM (PROCESS SAMPLING)

Sampling of particulate systems in a flowing or process stream is generally thought to provide more representative samples than from a stationary heap. However, care should still be taken to ensure samples are representative of the entire population. In addition to the constitutional and distributional heterogeneity of the bulk population, variations may also arise from time-based (temporal) or process-based factors. Examples of temporal factors may include shift changes in operators or equipment periodic maintenance. Examples of process-related factors may include fluctuations in temperature, pressure, or flow rate.

Three options exist when sampling from a flowing stream: (1) sample the whole stream during a fraction of the time, (2) sample a fraction of the stream for the whole time, or (3) sample a fraction of the stream for a fraction of the time. Numerous authors have stated that sampling should be conducted from a moving stream and that the whole stream should be for a fraction of the time rather than parts of the stream (3, 8, 9, 21). Sampling of a fraction of the stream may produce a biased sample which may not accurately reflect the entire stream's characteristics.

Incremental sampling is used to obtain particulate samples from a flowing stream. Analogous to incremental sampling from a stationary heap, multiple samples are obtained from different locations. For sampling from a flowing stream, multiple samples are collected at different time intervals from a process stream. Incremental sampling methods include random, systematic, stratified, and cluster sampling (Figure 4.2). These sampling approaches are probabilistic in nature.

Random sampling involves the selection of individual samples at random, without any specific pattern. Random sampling may be performed by assigning a unique identification number to individual objects in the population and choosing the sample by randomly picking numbers or using a table of random numbers (2). Random

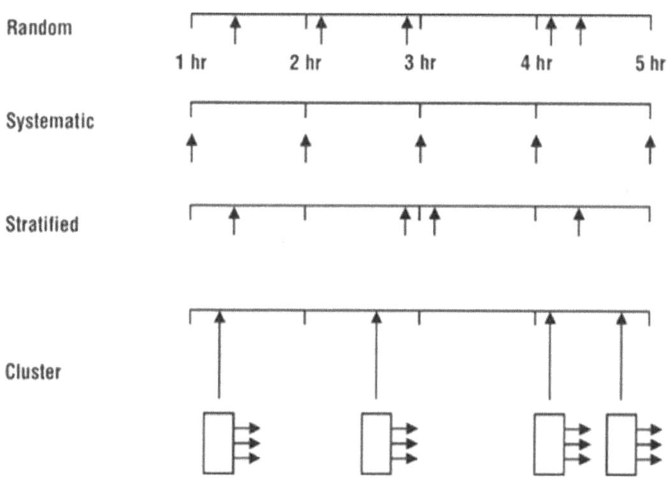

FIGURE 4.2 Illustration of incremental sampling schemes.

sampling is a common method of selecting samples. Random sampling is less reproducible than the other schemes and may produce an undesirable allocation.

Systematic sampling involves the collection of individual samples at specified intervals, with the initial sample being selected randomly (2). An example of systematic sampling is the collection of tablets at a particular time each hour during a production run. Systematic sampling is the most used sampling scheme. It is very reproducible, easy, and convenient to implement. However, using a systematic sampling scheme in a population with cyclic or periodic behavior poses a potential risk of significant error and biased results. Periodic functions exhibit a regular pattern of alternating minima and maxima (3–7). Although the sampling time interval is most unlikely to be an exact multiple of the periodic time interval associated with a particular bias, a stratified sampling scheme (described below) is recommended in the presence of periodic functions. Both systematic and stratified sampling methods generally provide the most reproducible selection.

Stratified sampling involves dividing the population into subsets (strata), and then samples are selected at random from within each stratum. The random sampling of 10 tablets during each hour from a production run is an example of stratified sampling. Stratified sampling is easily implemented and reproducible, and it suppresses the inherent error caused by periodic functions (3–7). However, data analysis with this method is complicated because it should account for stratification. Stratified sampling is recommended when the strata are very different from each other (i.e., there is large between-strata variability), but the individual units within each stratum are similar (i.e., there is small within-strata variability) (2).

Cluster sampling, also known as multistage sampling, is employed when many individual units are grouped together in larger clusters that may be subsampled. Clusters may be distinguished from strata in that they are discrete entities that may be sampled in their entirety. In single-stage cluster sampling, clusters are selected at random and all objects within the cluster are included in the sample. In two-stage cluster sampling, subsample units are chosen from the primary cluster (2). For example, 100 bottles are randomly selected from a production batch of 10,000 bottles, and 10 tablets are randomly selected from 100 tablets contained in each bottle. In this example, the bottle is known as the primary unit, the tablets are the subsample units, and the population is the production batch of tablets.

SAMPLE SPLITTING

Sampling may occur in a stage-wise process, where primary sampling involves obtaining samples from the bulk population and secondary sampling is required to divide the sample into smaller portions for testing purposes (21). The secondary sampling process is known as sample splitting. Sample splitting is the process of fractionating or dividing a single sample into smaller sub-samples. It may be used to generate replicate samples where multiple tests are conducted on the same original sample. It reduces the physical sample size from larger samples to laboratory or measurement-sized samples. Care should be taken to ensure the divided subsample accurately represents the original sample.

Sample splitting methods include fractional scooping, coning and quartering, table sampling, chute sampling, and spinning riffling (3, 8, 9, 21, 24, 25).

Fractional scooping separates the entire sample lot by manual or mechanical scooping. The sample retained is one scoopful out of a total of N, enabling a splitting ratio of 1/N. Three types of fractional scooping include true fractional scooping, degenerate fractional scooping, and alternate scooping. In true fractional scooping, the scoopfuls are deposited into N distinct heaps. Following the complete dealing out of the whole batch, the sample heap is selected at random. In degenerate fractional scooping, as the scoopfuls are dealt out, every Nth scoopful is deposited in the sample heap. The remaining (N - 1) scoopfuls of the cycle are deposited in the reject heap. In alternate scooping, the whole batch is divided into two heaps, and the sample heap is chosen at random following the completion of the scooping.

Coning and quartering, also known as "Cornish quartering," is the oldest probability sampling method, dating back to the early nineteenth century when it was used in the tin mines in Cornwall. Coning and quartering rely on the radial symmetry of the conical heap to produce four identical samples. The sample is poured into a conical heap from the central apex. This cone is flattened into a flat circular cake, retaining the symmetry achieved in the first step. The flattened powder bed is divided into four quarters by using a cross-shaped metal cutter. Two opposite quarters are retained, and the other two quarters are rejected. Reliable results are obtained if the heap is symmetrical about a vertical axis and the cutting planes align along the axis. However, in practice, the heap is unlikely to be truly symmetrical, resulting in different-sized samples. This method is very time-consuming and subject to operator bias; therefore, it is not recommended. It is no more accurate or precise than alternate shoveling.

Table sampling fractionates powders using a series of holes and prisms that are placed in the path of the powder stream. The powder is fed onto an inclined plane and is directed by strategically located prisms toward the holes or the sampling chute at the end of the plane. The sample is taken from the end of the plane, and the powder that falls through the holes is discarded. The accuracy of this method depends on the uniformity of the initial feed. Errors are further compounded by the sequential removal of parts of the stream, resulting in low accuracy.

The chute splitter consists of an assembly comprising an even number of identical adjacent chutes, each of which leads the powder alternately toward two different collection buckets. The sample is poured into the top of the chute and repeatedly halved until the desired sample size is achieved. Chute splitters are of various types, including a V-shaped trough divider, oscillating hopper divider, and oscillating paddle divider. This method is inexpensive and easy to use, and it gives accurate sample division when used properly. However, it is subject to operator bias. Potential bias may be suppressed by alternating the bucket chosen for sample retention. Overflow of one of the chutes may occur if material is discharged too quickly or close to one of the chutes.

The spinning riffler fractionates the powder into revolving collection containers under a stationary feeder. The powder should be filled into a mass flow hopper (steep sides at an angle greater than 70°) with minimal segregation. This is achieved by moving the pour point to prevent heap formation. The powder flows from the hopper onto a vibratory feeder and then finally into collection containers on a revolving table. The powder flow rate and collection stage revolutions should be maintained at a uniform speed to ensure similar fraction weights. Each container should receive more than 30 increments. Optimum results are obtained by using at least 35 collection containers. This method was found to be superior when compared to other methods. However, this method can be used only for free-flowing powders.

SAMPLING OF PHARMACEUTICAL PARTICULATE SYSTEMS

Pharmaceutical particulate systems may be sampled for characterization testing during the product development stage and commercial production. Samples may be taken from raw materials, intermediates, in-process samples, or finished products. Raw materials may be sampled for testing of identity, purity, impurities, and other physicochemical properties (e.g., polymorphic state, particle size, or moisture content). Intermediates or in-process samples may be taken to evaluate the unit process conditions. Some examples include monitoring the impurity levels of the API intermediate to determine if further purification is required, assessment of the blend content uniformity to determine whether longer mixing time is needed, evaluation of moisture content of granules to determine whether longer drying time is needed, or in-process testing of tablet hardness, friability, disintegration, and dissolution to assess whether compression forces need to be adjusted. Finished products are sampled and tested against the product specifications to ensure acceptable product quality.

During product development, the effect of changes in the input materials and process parameters on the physicochemical properties of the intermediates and the finished product is evaluated. The level of sampling and testing conducted during the R&D phase is generally greater than during commercial production. Characterization of the physicochemical properties of the input materials (API and excipients), intermediates, and finished products is needed for formulation and manufacturing design and optimization. It is important that representative samples are obtained to ensure accurate data is available for decision-making.

Good Manufacturing Practices (GMPs) need to be followed for batches manufactured and tested for clinical studies, formal stability, process validation, and commercial production. This includes the use of written sampling plans that describe the number of samples taken, the sampling method (e.g., composite and stratified), sampling locations or time points, and tests to be conducted. The details of the sampling plans should be justified. Sampling plans are required for raw materials, in-process or intermediate products, packaging materials, and finished products. The sampling plan should include details of the material type being sampled, the number of samples and sampling location (e.g., sampling diagram), the time and frequency of sampling (e.g., every 30 minutes), the sampling procedure, the sample quantity, sampling equipment or tools, and any safety or environmental considerations (24). The sampling plan and details of the samples taken are documented in the manufacturing and packaging batch records. During process validation, an enhanced level of sampling may be used to verify that the proposed manufacturing process produces intermediate and finished products with the desired characteristics and complies with specification limits. Following process validation, the level of sampling may be reduced for commercial production, especially for in-process and intermediate product testing.

The sampling of pharmaceutical particulate systems follows Gy's foundational principles, where samples may be from stationary piles or flowing streams (also known as process sampling) (3–7). Generally, raw materials are sampled from their storage containers (stationary heaps) during incoming testing. In-process and intermediate products may be sampled from stationary heaps or as flowing streams, depending on the unit operation and whether sampling is accessible from the manufacturing equipment. Process sampling from some unit operations is not feasible. The product from micronization or spray-drying processes is accumulated in a cyclone system; thus, samples must be taken after completion of the process. Some samples may be taken directly from the manufacturing equipment at planned intervals while the process is in progress. For example, during product development, the blending operation may be stopped for sampling to assess blend homogeneity. Samples are taken directly from the blender using a sampling thief at specific time intervals. The blending speed and time are selected based on the test results. Other examples include samples obtained from a fluid bed dryer to assess whether the target moisture content has been reached or samples taken from a tablet coating machine to assess whether the target tablet coating thickness (weight) has been reached. In those cases, the manufacturing process is stopped, and a sample is taken and tested. If the target criteria are reached, the process is stopped. If the target criteria have not been reached, the process is continued until the target is reached. In contrast, for some manufacturing processes, the product can be sampled from the exit port while the process is ongoing (i.e., flowing stream). Examples include sampling of hot melt extrusion, dry granulation, or tablet compression.

Drug content uniformity is a critical attribute that ensures that each dosage unit contains the required amount of drug. During drug development, content uniformity testing is performed on intermediate products to ensure the manufacturing process provides acceptable drug homogeneity (e.g., engineered particles, granules, and blends). Acceptable content uniformity results from earlier unit operations provide a greater level of confidence that the finished product will also have acceptable content uniformity. In-process test results allow for problem-solving in cases where the finished product is shown to have content uniformity issues. Ideally, it is useful to have the in-process results of intermediate products prior to manufacturing the finished product. However, sometimes samples of the intermediate products are taken but testing is either conducted in parallel or only after the finished product results are obtained.

Blend uniformity testing is a common type of drug content uniformity test performed during product development. It determines whether the active ingredient is homogeneously dispersed throughout the powder blend. Blend samples are usually taken from the blender or an intermediate storage container using a sampling thief. The sampling locations should be selected to provide a representative cross-section of the blend and should include areas with a high potential for non-uniformity. The sample weight of blend samples is required to be no more than three times larger than the intended unit dose of the finished product. This is a legal requirement based on the Barr decision of February 1993 (*U.S. District Court for the District of New Jersey vs. Barr Laboratories, Inc.*, Civil Action No. 92-1744). The premise of this court decision is that the scale of scrutiny should match that of the final dosage form. General recommendations for blend uniformity include obtaining three replicate samples at 10 locations in the powder blend. The 30 replicate samples should be individually analyzed, and the results should be evaluated to determine the between-location and within-location variability. High variance in the between-location samples is an indicator of inadequate mixing to produce a homogeneous blend. High variance in the within-location samples is an indicator of either sampling bias or incomplete mixing at the unit-dose level (26). The US FDA has stated that although powder thieves are known to have limitations, they are still widely used and provide reliable results in many cases (27). Blend uniformity testing is useful for blending process design and optimization during product development. Acceptable blend uniformity improves the chances of meeting content uniformity test requirements for the finished product. However, this is not guaranteed due to potential segregation during storage, handling, and onward processing.

Content uniformity of dosage units is generally tested on the finished drug product during commercial production according to pharmacopeial monographs, such as the harmonized USP <905> and Ph. Eur. 2.9.40 monographs. These monographs provide the methodology and acceptance criteria for testing the uniformity of dosage units for solid, semi-solid, and liquid dosage forms, including tablets, capsules, solids, or solutions in single-dose containers, and others. Uniformity of dosage units can be demonstrated by either content uniformity or mass variation. The mass variation method is limited to specific cases: solutions in single-dose containers, single component solids in single-dose containers (e.g., no inactive ingredients), solids prepared from solutions and freeze-dried, and tablets or capsules containing 25 mg or more active substance and comprising of 25% or more by mass of the dosage unit. Content uniformity is required for all other dosage forms. Thirty units are selected from the batch, and 10 units are individually analyzed. The Acceptance Value (AV) is calculated. The batch meets the acceptance criteria if the calculated AV result is not more than 15. If the AV result from the initial 10 units is greater than 15, then the remaining 20 units are analyzed. The batch meets the acceptance criteria if the calculated AV result from the 30 units is not more than 15.

Content uniformity of dosage units with enhanced sampling may occur during process validation to provide an understanding of the manufacturing process and to provide confidence that the USP <905> or Ph. Eur. 2.9.40 criteria will be met during routine production. A commonly used methodology and acceptance criteria are described in ASTM E2709 and E2810 standards, respectively. This is a statistical-based approach that provides a 95% confidence level and 95% coverage (i.e., probability of passing) with respect to the USP <905> Uniformity of Dosage Units monograph. The samples are taken at multiple locations throughout the manufacturing run from the beginning to the end in regular intervals using a stratified or systematic sampling procedure. Sampling Plan 1 involves a single dosage unit taken at each location. Sampling Plan 2 involves multiple dosage units taken at each location, which allows the determination of within- and between-location variability. The samples are collected and analyzed. The acceptance limits are provided in tabular format in ASTM E2810. For Sampling Plan 1, the acceptance limits provided are the upper bound limit of the standard deviation, based on the sample size used and the sample mean observed. If the observed standard deviation falls below the upper limit, then the batch results are acceptable. For Sampling Plan 2, the acceptance limits provided are the lower and upper limits for the sample mean, based on the number of sample locations, the number of replicates at each location, the pooled within-location standard deviation, and the standard deviation of the location means. If the sample mean falls within the acceptance limits, then the batch results are acceptable, which means that there is a 95% confidence level and a 95% chance of meeting the USP <905> criteria.

Stratified sampling may be employed for other release tests of the finished products. For drug performance tests, such as dissolution or aerodynamic particle size distribution, samples are commonly collected at the start, middle, and end of the manufacturing run. The stratified testing provides supportive data to demonstrate that the drug product performance is consistent throughout the entire manufacturing run.

SAMPLING, PREPARATION, AND MEASUREMENT ERRORS

According to Gy's principles, the global estimation of error is the sum of total sampling error and total analytical (measurement) error (3–7). If it is assumed that sample preparation and measurement are performed without error, the difference between a sample characteristic and its corresponding population parameter would be due to sampling error. Ideal homogenous systems (e.g., miscible solutions) do not exhibit any sampling errors. However, all particulate systems exhibit some degree of constitutional and distributional heterogeneity and, thus, are susceptible to sampling errors.

The total sampling error (TSE) can be divided into three components: correct sampling errors (CSE), point selection errors (PSE), and incorrect sampling errors (ISE). These components are further divided into sub-components.

Correct sampling errors (CSE) are caused by the constitutional heterogeneity (known as the fundamental sampling error, FSE) and distributional heterogeneity (known as the grouping and segregation error, GSE) of the bulk material. If the sampling process is performed correctly, these factors provide the true variability of the bulk material.

Point selection errors (PSE) may occur when sampling from a flowing stream (process sampling) and are related to time-based variations in the process, which may be cyclic or non-cyclic. Potential sources of time-based variations include flow-rate or pressure-related fluctuations in the manufacturing equipment or machine stoppages due to operator shifts.

Incorrect sampling errors (ISE) may consist of three sub-components: (1) incorrect delimitation error (IDE), which is related to the geometric location of the sample;

(2) incorrect extraction error (IEE), which is related to the removal of the sample from the bulk material; and (3) incorrect processing error (IPE), which is related to the handling of the extracted sample. An example of IDE is when a metal cutter is not placed symmetrically on the vertical axis during coning and quartering, resulting in unequal quarters of powder. An example of IEE is when a delimited sample is not fully removed from the bulk material. This may occur using a sample thief where particles are not fully captured. Examples of IPEs include loss or contamination of the sample material after extraction from the bulk material, alterations in the physical or chemical composition of the sample, and deliberate or involuntary operator faults. As the name suggests, incorrect sampling errors are related to situations when the sampling process is not performed correctly.

Chemical or physical changes may occur to the sample in the period between sampling and analysis. Oxidation or hydrolysis may alter the chemical composition of the sample. Moisture adsorption or loss may occur. Extreme temperature changes may impact the chemical or physical composition of the sample. Vibrational segregation may occur during transport. Particle comminution may occur if samples are roughly handled, potentially impacting the particle size and shape. These examples highlight the importance of appropriate packaging, transport, and storage conditions of the samples to maintain the sample integrity prior to analysis. For example, samples may be packaged in LDPE double bags, and then packaged with a desiccant inside a sealed aluminum foil pouch to prevent moisture ingress.

Analytical errors may occur during sample preparation, analysis, or measurement. Sources of variation include the sample, instrumentation, operator, and background noise, which includes such environmental variables as temperature and humidity (2). Samples received in the analytical laboratory may need to be further sampled to obtain the amount required for the analysis or measurement. The sample quantity needed for measurement varies depending on the analytical method or measurement technique. A common approach to withdraw the quantity required for analysis from the laboratory sample is to measure out the appropriate amount using a spatula. Sometimes the sample container may be shaken or stirred to attempt to mix the particles before sampling. However, this approach is subject to segregation and bias. The sample removed by a spatula includes particles from the surface, which may be different in composition than the bulk powder. Also, sample removal using a spatula may result in size segregation, where the finer particles are captured, and coarser particles are lost due to flowing down the sloping surface. This effect is especially common when using a flat spatula. Instead, sample splitting methods (described above) are recommended for the reduction of laboratory-sized samples to measurement-sized samples (8).

Measurements should demonstrate acceptable levels of accuracy and precision. Accuracy refers to how close the measured value is to the true value. Accurate measurements are free from systematic error and bias. This is visually represented by how close the arrow hits the target. The target may be determined by reference standards. Precision refers to the reproducibility of repeated measurements using the same conditions. This is visually represented by the scatter of arrows around the mean value. Precision may be described using the standard deviation, relative standard deviation (RSD), or the coefficient of variation (CV). Method accuracy and precision are assessed during method validation. Depending on the method, other validation parameters may also be evaluated, such as specificity, linearity, range, limit of detection, limit of quantification, and robustness.

Correct measurements require both accuracy and precision. This ensures the result is close to the true value and consistent over repeated measurements. Measurement errors occur when the measurement is accurate but not precise, precise but not accurate, or neither accurate nor precise. If the measurement is accurate but not precise, this means that results are close to the true value, but there is high variability. If the measurement is precise, but not accurate, this means the results are tightly clustered but deviate from the true value, often caused by systematic bias. If the measurement is neither accurate nor precise, this means that the results are widely scattered around a value that deviates from the true value.

Measurement errors may occur due to sample, operator, or instrument-related errors (intentional or unintentional mistakes). The following section illustrates preparation and measurement errors using particle size measurement as an example.

PREPARATION AND MEASUREMENT ERRORS IN PARTICLE SIZE MEASUREMENTS

Particle size measurements are subject to sampling, preparation, and measurement errors. Each particle size measurement technique uses a distinct measurement principle. Thus, a particle size measurement from one sizing technique is expected to be different from those obtained from a different sizing technique for the same material sample.

Particle sizing instruments determine equivalent spherical diameters based on different principles (28) (covered in more detail in Chapter 5). Particle size may be determined from measurements in one dimension (diameter, width, or length), two dimensions (area), or three dimensions (surface area or volume). For spherical particles, these estimates will be numerically similar. For non-spherical particles, these estimates vary greatly based on their particle shape. The size measurements for a given material will vary, depending on the measurement principle used. Also, measurement bias may occur if the sample size distribution falls outside the size range accurately measured by the instrument. This highlights the importance of understanding the measurement principle and limitations of each particle size instrument to ensure a meaningful interpretation of the measured result (29).

The degree of particle dispersion is another source of variability in particle size measurements. The measured particle size may be that of individual particles, aggregates, or flocculates. An aggregate is a group of two or more individual particles strongly held together by forces that are stable in normal handling but broken up by shear forces. A flocculate is a group of aggregates or particles in a liquid suspension that break up by normal shaking and stirring but rapidly reform on standing (28). Particulate samples may be dispersed in a dry or wet state. For dispersion of dry powders, shear forces are commonly provided by pressure drops between the shear forces applied and the inter-particulate forces within the aggregates. Care should be taken to prevent particle comminution. When particles are dispersed in a liquid medium, the dispersant should not produce dissolution or be chemically reactive with the sample. Dispersion may occur by stirring or sonication. Surfactants may be added to aid particle wetting and dispersion. Measurement bias may be a result of incomplete dispersion of aggregates and particle comminution.

Microscopy

Particle size measurements by microscopy are conducted by using a microscope to visualize particles using high magnification. The higher resolution enables a detailed examination of individual particles. There are many different microscopy techniques. Optical microscopy and scanning electron microscopy (SEM) are commonly used for particle size measurements. Optical microscopy uses visible light to illuminate and magnify samples. It is useful for particles in the micrometer to millimeter range. Sample preparation involves dispersing particles in a suitable mounting media to provide individual particles, uniformly distributed across the microscope slide. Particle size measurements can be performed by manual comparison to the size scale or using image analysis software from the captured images.

In contrast, SEM uses electron beams instead of light to visualize samples. This enables higher magnification and resolution. It is useful for particles in the nanometer to micrometer range. Sample preparation includes fixing the sample onto a metal stub and coating it with a thin layer of gold. This improves the image quality by enhancing the conductivity and reducing charging effects. The high-resolution SEM images allow particle size, shape, and surface morphology to be characterized. Particle size measurements can be performed by manual comparison to the size scale or using image analysis software from the captured images.

Measurement errors may occur due to incorrect equipment calibration or operator subjectivity due to skill, expertise, or judgment during manual measurements. Difficulties may arise due to poor image resolution or limited depth of field resulting in blurred images. If samples are not adequately dispersed into individual particles, then the size measurements may include aggregates. There may be unclear boundaries due to touching particles or aggregated clusters. There may also be artifacts from sample preparation or imaging quality.

Size measurements should be conducted on a series of randomly selected fields, and all particles in the field should be measured to avoid bias. Representative sampling requires measuring at least 300 particles. Measurement of insufficient sample size may be a source of error (28).

Sieving

Particle size measurements conducted by the sieving technique involve passing a sample through a series of sieves. The smaller particles pass through the sieve opening while the larger particles are retained on the top of each sieve, depending on the aperture size. The sieves are typically made of stainless steel with standard aperture sizes. The sieves are stacked with the largest aperture size on top, then decreasing aperture sizes below, and a collection pan under the bottom sieve to collect any material that passes through. The sample is introduced to the top sieve and covered with a lid. The sieve stack is placed on a mechanical shaker. After sieving is completed, the material retained on each sieve is carefully recovered and weighed. The particle size distribution is plotted on a histogram showing the amount or percentage of powder retained on each sieve with a defined size aperture. The results may be expressed as a percentage of powder above or below a specific size, or within a specific size range.

Method parameters for particle size measurement by sieving include the sieve loading, duration of sieving, frequency, and amplitude of sieving. The passage of a particle through a sieve aperture may not solely depend on its relative size dimensions. Some factors affecting particles passing through the sieve aperture include the size and quantity of surrounding particles, access to the aperture, deviations from the nominal size of the aperture, and the relative particle orientation compared to the aperture. Aggregation of small particles may prevent passage through the sieve aperture. Particle comminution may allow larger particles to pass through smaller apertures (28).

Sedimentation

Particle size measurements using the sedimentation method are based on Stokes' law, with the following assumptions: particles are rigid spheres, the fluid is incompressible, the fluid velocity at the particle surface is zero, and particle motion is constant and not affected by other particles or nearby walls. However, these assumptions are not always met. Corrections are required for non-spherical particles and the effects of the container bottom and walls. The motion of neighboring particles may affect the sedimentation rate and calculated particle size (28). Accurate size measurements from sedimentation methodology require complete particle dispersion in a suitable liquid media with suitable viscosity such that the particle sedimentation rate lies within the Stokes region. Temperature control may be needed to prevent changes in the viscosity of the dispersing media. A dispersing agent may be added, but the dispersing media should be free of air bubbles and disturbances (28).

Laser Diffraction

Laser diffraction instruments use either the Fraunhofer or Mie theories to determine particle size from the diffraction pattern of particles. The Fraunhofer theory assumes that the wavelength of light is significantly smaller than the particle diameter, particles of all sizes scatter light with equal efficiency, and the particle is opaque and transmits no light. However, these assumptions are not valid for many pharmaceutical particles. The Mie theory is more appropriate for particles smaller than 10 μm; however, it requires knowledge of the refractive indices of the particulate material and dispersing medium. Laser diffraction instruments may exhibit inaccurate measurements for small-sized or transparent particles. Also, small "ghost" particles may appear due to high-angle diffraction of the edges of non-spherical particles.

Laser diffraction instruments employ mathematical algorithms to determine the particle size distribution from the complex diffraction patterns obtained. These algorithms are generally considered proprietary and are not publicly disclosed. The algorithms differ between manufacturers and software versions, which can lead to significant differences in the particle size distribution results of the same sample using either Fraunhofer or Mie optical models (30).

CONCLUSION

Sampling is the process of selecting a small part of a population to estimate a given characteristic of the entire population. Representative sampling is critical for the estimation of the characteristics of the entire population. Representative samples are obtained when probabilistic sampling occurs, and sample integrity is maintained. Pharmaceutical particulate systems exhibit both constitutional (intrinsic variations) and distributional (spatial variations) heterogeneity. Particulate systems can be sampled from a stationary heap using grab, composite, or incremental sampling techniques, or from a moving stream using incremental techniques. Sampling from a moving stream is stated to produce more representative samples, especially when the whole stream is sampled during a fraction of the time. Incremental sampling methods include random, systematic, stratified, and cluster sampling. Sample splitting techniques may be used to further divide primary samples taken from the bulk material population into smaller portions for laboratory testing.

In the pharmaceutical industry, sampling plans document the number of samples, sampling method, sampling location, or time points to comply with Good Manufacturing Practices. Sampling plans are needed for raw materials, in-process or intermediate products, packaging materials, and finished products. Sampling for blend uniformity and uniformity of dosage units testing was described in depth. Sampling, preparation, and measurement errors were also discussed, and examples are provided for particle size measurements.

Knowledge of the characteristics and performance of powders is interpreted through the filter of sampling and statistical probabilities. The pharmaceutical scientist and powder technologist should always review the nature of this filter and remember that a true population parameter exists for which an estimate is all that can be achieved. The goal is to ensure that this estimate is a reliable reflection of the true parameter and is relevant to the desired application.

REFERENCES

1. Thompson SK. *Sampling*. New York: Wiley; 1992.
2. Bolton S (ed). *Pharmaceutical Statistics: Practical and Clinical Applications*. New York: Marcel Dekker; 1997.
3. Gy PM. *Sampling of Particulate Materials: Theory and Practice*. Amsterdam, The Netherlands: Elsevier; 1982.
4. Gy P. Sampling of discrete materials - a new introduction to the theory of sampling I. Qualitative approach. *Chemom Intell Lab Syst*. 2004;74:7–24.
5. Gy P. Sampling of discrete materials - a new introduction to the theory of sampling II. Quantitative approach. *Chemom Intell Lab Syst*. 2004;74:25–38.

6. Gy P. Sampling of discrete materials - a new introduction to the theory of sampling III. Quantitative approach - sampling of one-dimensional objects. *Chemom Intell Lab Syst.* 2004;74:39–47.
7. Gy P. Sampling of discrete materials - a new introduction to the theory of sampling IV. 50 years of sampling theory. *Chemom Intell Lab Syst.* 2004;74:49–60.
8. Allen T. *Particle Size Measurement.* London: Chapman and Hall; 1990.
9. Brittain HG. The problem of sampling powdered solids. *Pharm Technol.* 2002:67–73.
10. Chowhan ZT. Segregation of particulate solids, Part I. *Pharm Tech.* 1995;5:56–70.
11. Chowhan ZT. Segregation of particulate solids, Part II. *Pharm Tech.* 1995;6:80–94.
12. Rhodes M. Mixing and segregation. In: Rhodes M, editor. *Introduction to Particle Technology.* Chichester, UK: Wiley; 1998. p. 223–40.
13. Venables HJ, Wells JI. Powder mixing. *Drug Dev Ind Pharm.* 2001;27:599–612.
14. Esbensen HK, Romanach RJ, Romano-Ospino AD. Theory of sampling (TOS): A necessary and sufficient guarantee for reliable multivariate analysis in pharmaceutical manufacturing. In: Tobyn M, Menezes JC, Ferreira AP, editors. *Multivariate Analysis in the Pharmaceutical Industry.* New York: Academic Press; 2018. p. 53–91.
15. Carstensen JT, Dali MV. Blending validation and content uniformity of low-content noncohesive powder blends. *Drug Dev Ind Pharm.* 1996;22:285–90.
16. Muzzio FJ, Robinson P, Wightman C, Brone D. Sampling practices in powder blending. *Int J Pharm.* 1997;155:153–78.
17. Mohan S, Rankell A, Rehm C, Bhalani V, Kulkarni A. Unit-dose sampling and blend content uniformity testing. *Pharm Tech.* 1997;21:116–25.
18. Berman J, Schoeneman A, Shelton JT. Unit dose sampling: A tale of two thieves. *Drug Dev Ind Pharm.* 1996;22:1121–32.
19. Garcia TP, Taylor MK, Pande GS. Comparison of the performance of two sample thieves for the determination of the content uniformity of a powder blend. *Pharm Dev Technol.* 1998;3:7–12.
20. Muzzio FJ, Goodridge CL, Alexander A, Arratia P, Yang H, Suddah O, et al. Sampling and characterization of pharmaceutical powders and granular blends. *Int J Pharm.* 2003;250:51–64.
21. Venables HJ, Wells JI. Powder sampling. *Drug Dev Ind Pharm.* 2002;28:107–17.
22. Muzzio FJ, Roddy M, Brone D, Alexander AW, Suddah O. An improved powder-sampling tool. *Pharm Tech.* 1999;23:92–110.
23. Susana L, Canu P, Santomaso AC. Development and characterization of a new thief sampling device for cohesive powders. *Int J Pharm.* 2011;416:260–7.
24. Smith PL. Audit and assessment of sampling systems. *Chemom Intell Lab Syst.* 2004;74:225–30.
25. Peterson L, Dahl CK, Esbensen KH. Representative mass reduction in sampling - A critical survey of techniques and hardware. *Chemom Intell Lab Syst.* 2004;74:95–114.
26. Bergum JS, Prescott JK, Tejwani RW, Garcia TP, Clark J, Brown W. Current events in blend and content uniformity. *Pharm Eng.* 2014;34:1–10.
27. US FDA Questions and &Answers on Current Good Manufacturing Practice. Regulations - Production and Process Controls, Question 15; https://www.fda.gov/drugs/guidances-drugs/questions-and-answers-current-good-manufacturing-practice-regulations-production-and-process#15
28. Herdan G, Smith ML, Hardwick WH, Connor P. *Small Particle Statistics.* London: Butterworth-Heinemann; 1960.
29. Etzler FM, Sanderson MS. Particle size analysis: A comparative study of various methods. *Part Part Syst Char.* 1995;12:217–24.
30. Etzler F, Deanne R. Particle size analysis: A comparative study of various methods II. *Part Part Syst Char.* 1997;12:278–82.

5 Particle Characterization and Data Expression

The following sections describe characteristics of particles that are critical to their performance and the quality of products in which they are incorporated. The principal characteristics and the means of assessing their quality and performance are described in sufficient detail to establish the relationship of certain properties, for example, particle size and surface area, and the complementary nature of measurements that are made to establish dimensions and properties that contribute to the behavior of particles and powders.

PARTICLE SIZE

The principles behind various methods employed to assess particle morphology are embodied in myriad commercially available instruments. Their operation is illustrated in this chapter, although details of particular instruments will not be described. Particle size and morphology measurement may be divided into four classes: direct imaging, indirect imaging, physical methods, and charge analysis. A number of different instruments lie within these broad categories, each with its own particular strengths and limitations. Consideration of approaches to calibration of particle sizing instrumentation is also presented. Finally, comments are made on the selection criteria for a particular method or group of methods. The average dimension being measured is the most important criterion for choosing any instrument. Figure 5.1 presents a sketch of the operational range of particle sizes for the methods described in this chapter. The significant overlap among the techniques provides the opportunity to utilize more than one method to develop a more complete picture of the overall product.

DIRECT IMAGING

Direct imaging is an absolute method of particle size and morphology characterization because it is the only technique in which the instrument itself does not provide some interpretation of the data. Direct imaging is also the only method that provides the ability to examine individual particles rather than a population of particles. Microscopy, a direct-imaging technique, provides a confirmation for any other method of particle sizing. It also provides the ability to identify irregular particle morphology as well as aggregates, hollow particles, or other forms of non-homogeneity. Sizing particles in a microscope image involves comparing the particles with a reference, either in real time or from a captured image. Therefore, some interpretation is required in direct imaging because the reference comparison is operator dependent. In addition, for nonspherical particles, using various equivalent diameter

DOI: 10.1201/9781003470151-5

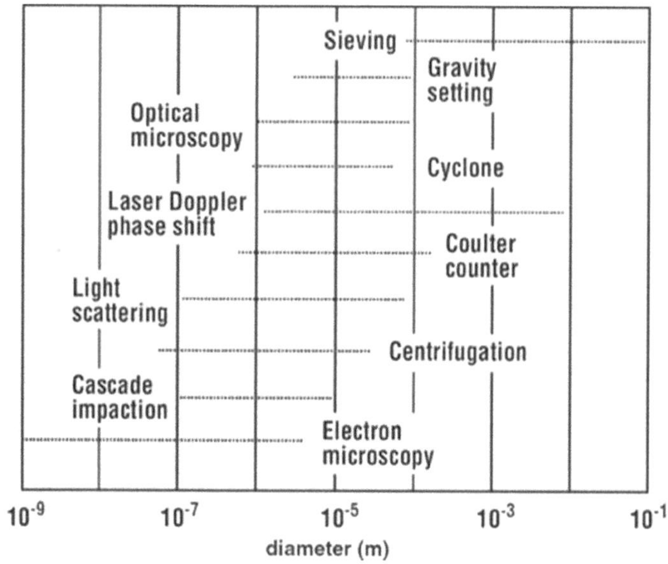

FIGURE 5.1 Range of operation for particle sizing methods.

expressions, as described later in this chapter, introduces further errors of interpretation. Direct imaging includes the methods of optical, electron, and atomic force microscopy, as described in the following sections.

Optical Microscopy

Optical microscopy uses visible light to examine particles. The wavelength of light determines the fundamental limit of resolution for this method. The resolving power of a microscope is equal to the wavelength of illuminating light divided by the sum of the numerical aperture of the objective lens and the numerical aperture of the condenser. The numerical aperture is determined by the index of refraction between the object and the lens and the half-angle of light entering the lens. There is a trade-off, however, between resolution and depth of focus: increasing the numerical aperture increases resolution but decreases the depth of focus. In general, the highest focal length (depth of focus) suitable for a given sample should be employed. Optical microscopy can, in theory, be used to image particles larger than about 1 μm. Practically, however, difficulties in the depth of focus at high magnifications limit the utility of light microscopy for particles sized near the resolution limit. Electron microscopy, discussed in a later section, is the method of choice for sizing of particles near, and somewhat above, the light microscopy resolution limit.

Microscopy requires the preparation of a slide containing a representative sample of uniformly dispersed powder. Most methods of preparing the slide involve dispersing the sample as a suspension in a liquid and then spreading a drop of the fluid on the slide (1). The result of this procedure is, unfortunately, largely operator dependent. The powder-containing fluid can be spread either by centrifugal force or by physical means such as a glass rod. Care must be taken when spreading powders to not fracture the particles. The goal in spreading the particles is to achieve a low density of particles per unit area (2). High densities can result in judgment errors in sizing because the size of nearby particles in the image may influence the size determination made by the operator.

Determination of particle sizes from microscope images may occur in real time or from a captured image. In both cases, a reference scale placed in the image is compared to particles in the image. An older method of measuring particle sizes uses a microscope equipped with an optical micrometer linked to a movable crosshair in the ocular. The crosshair is moved to the edge of a particle, the micrometer reading is recorded, and the crosshair is then moved to the opposite edge. The difference in micrometer readings between opposite edges is the particle diameter. This is clearly a very time-consuming technique and has been supplanted by other methods.

Particles may be sized by ocular scales or graticules. In each case, the slide is placed on a stage that can be moved in perpendicular directions. The slide is examined in strips, and the size of the particles passing over the scale in the image is recorded. Measurement is performed in vertical or horizontal strips to provide thorough coverage of the slide. The diameter of nonspherical particles is taken in the direction of the scan with the idea that differences in the orientation of particles will average out over the whole image. Particle size determination by graticules involves the comparison of individual particles with circles of decreasing diameter. Most graticules use a $\sqrt{2}^{n}$ progression in circle diameter, where n is the nth circle. This progression is used because many aerosol systems studied with these graticules are distributed lognormally, and the $\sqrt{2}$ progression gives equally spaced data points when plotted on logarithmic graph paper. The Patterson Globe and Circle or Porton graticules are examples of this type of graticule. Graticules should be calibrated before use with a stage micrometer.

Achieving a representative sample of the overall particle size distribution clearly requires sampling a large number of particles in the image. Various sources recommend differing counts suggesting that as few as 600 particles may be counted (1). The ASTM method requires the measurement of 10 particles in the largest and smallest size classes in the tails of the distribution (3). This may not be practical for a very narrow particle size distribution because, statistically, so few particles would be found in the tails that very large sample sizes would be required.

Microscope images may be acquired digitally and sized with an automated or semi-automated approach. It should be noted that the combination of photography and optical imaging could lead to errors in measurement because weight is given to particles in focus. In practice, a number of fields of view should be used for statistical validity. In the semiautomated method, an operator uses the computer mouse to draw reference lines across the particles. The software calculates the diameters from the length of a calibration line in the image whose length is entered by the user. Digital acquisition offers the further advantage of readily available archiving and reproduction of the image data. Automated systems are also available. There is a potential for error in these systems, however, especially if the distribution of powder in the image is too dense. The software may view two overlapping particles as a single particle. The semiautomated method combines ease of use with the ability to include human judgment in the process.

Optical microscopy is the least expensive and most readily available method of particle sizing. Drawbacks to the microscope include the potential for operator error, especially given the tedium of particle sizing by this method. Additionally, for a significant statistical representation of a size distribution, at least 300 particles should be counted. Finally, this method does not offer the ability to separate particles by size.

Polarized Optical Microscopy

Polarized light microscopes add two polarizing filters to an ordinary optical microscope—one below the microscope stage to polarize illuminating light and one situated between the objective lens and the eyepiece. In addition to the polarizing filters, the microscope should be fitted with a rotating sample stage and a crosswire graticule to mark the center of the field of view. Polarized light microscopy provides the ability to distinguish between isotropic and anisotropic materials. Isotropic materials demonstrate the same optical properties in all directions, whereas anisotropic materials have optical properties that vary with the orientation of incident light with the crystal axes.

In the typical setup, the two polarizing filters are oriented with their polarization directions at right angles so that, in the absence of a sample, no light passes through the system and the field of view is dark. A sample placed on the stage is rotated through 360 degrees. With the crossed polarizers, anisotropic materials can be distinguished from isotropic materials because they will be illuminated at some angle during rotation, whereas the isotropic materials will remain dark. In another use, the analyzer polarizer can be rotated out of the optical path. Anisotropic materials will change color when rotated because their absorption color varies with the orientation of the crystal structure to the polarized incident light. Polarizing microscopy has been used to examine the crystal habit and particle size of crystalline drugs (4).

Dark-Field Microscopy

As with light microscopy, dark-field microscopy creates a contrast between the object being imaged and the surrounding field. However, as the name implies, in dark-field microscopy the image background is dark and the object is bright. This technique is particularly useful for visualizing samples that have refractive indices very close to those of their surroundings (5).

In conventional light microscopy, an image of the sample is formed due to the reflection of light off the sample or due to the refraction of light from changes in the refractive index as light strikes the sample. Both reflection and refraction produce small angular changes in the direction of the light ray. In the case of a sample with a low index of refraction, the angular changes are so small as to be overwhelmed by the unrefracted light from the illumination source. The dark field is produced by an

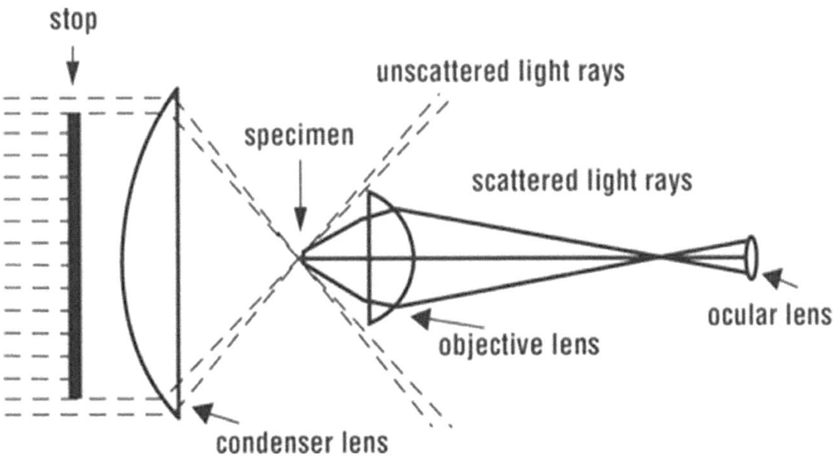

FIGURE 5.2 Dark-field microscopy setup.

annular stop placed in the path of the illuminating light before it reaches the specimen (Gray 1964). Almost any laboratory microscope can be converted to a dark-field microscope by adding an opaque stop. With the stop in place, only oblique rays pass through the specimen on the microscope slide (see Figure 5.2). With no specimen, an oblique hollow light cone is formed in the direction of the objective lens. If the numerical aperture of the condenser is greater than the objective, the oblique rays will fail to enter the objective and the field will appear dark. Reflection, refraction, and diffraction from discontinuities in the sample will alter the path of incident light so that it is collected by the objective. The result is an illuminated image in a dark field. In Fourier terms, the annular stop serves to remove the zero order (i.e., the unscattered light) from the diffraction pattern, leaving only higher-order diffraction intensities to be collected by the objective.

Impurities such as dust will scatter light as effectively as the sample, so much care must be taken when preparing slides for dark-field microscopy. Specimen thickness is an important consideration. A thin sample is preferable to minimize diffraction artifacts that can interfere with the image. Microscope slides will have the same effect and therefore should be no more than 1 mm thick. Because the annular stop eliminates bright, undeviated light, a high-intensity light source must be used (6).

Fluorescence Microscopy

Fluorescence refers to the reemission of absorbed light at longer wavelengths than the excitation light. This occurs when a photon of excitation radiation collides with an electron in the atomic cloud, exciting it to a higher energy level. The excited electron then relaxes to a lower level and, because energy must be conserved, emits the light as a lower-energy photon. Fluorescence microscopy takes advantage of this effect to image materials that fluoresce either naturally or when treated with fluorescing chemicals. Fluorescence microscopes excite the sample with ultraviolet

(UV) light and observe emitted light in the visible spectrum. A filter is required in the objective of the microscope to block UV light reflected by the sample and pass only the fluorescence component. Fluorescent radiation is spherically symmetrically emitted by the fluorophore. The method is very sensitive; as few as 50 fluorescing molecules per cubic micron can be detected.

The technique can be applied to inorganic or organic—including living—material (7, 8). For secondary fluorescence (i.e., samples treated with fluorescing chemicals), a number of fluorochromes are highly specific in their attachment targeting. A commonly used fluorochrome is fluorescein-isothiocyanate (FITC), a molecule that is effective for marking proteins. It is widely used in a fluorescent antibody technique for identification of pathogens. An interesting application of this microscopy is the visualization of particle uptake by macrophages (9). Fluorescence microscopy provides the ability to view this process in real time.

Transmission Electron Microscopy

The fundamental limit of resolution imposed by the wavelength of light prompted the development of the transmission electron microscope (TEM). Decreasing the wavelength through the electromagnetic spectrum would suggest the use of UV or X-ray wavelengths for microscopy. Unfortunately, UV light is absorbed by glass, so UV microscope optics must be constructed of quartz. X-ray microscopes have been constructed but require extremely intense sources, such as synchrotrons, and complicated optics, such as Fresnel zone plates (10).

The wave/particle duality of electrons provided an alternative source. Electrons can be accelerated by an electrical potential and refracted in a magnetic field. The faster the velocity of the electron, the shorter the wavelength of the electron wave, as given by the de Broglie equation:

$$\lambda = \frac{h}{mv} \tag{5.1}$$

where λ is the wavelength, h is Planck's constant, m is the electron mass, and v is the velocity. Transmission electron microscopes offer approximately a thousandfold increase in resolution and a hundredfold increase in depth of field.

A drawback of the TEM is that electrons are quickly absorbed by any matter they encounter. Therefore, TEM imaging must occur under a high vacuum to minimize electron collisions with air molecules. The sample being imaged must be sectioned extremely thin for the electrons to penetrate the material. Finally, the TEM stage must be scanned through the electron beam. The scanning electron microscope (SEM) addresses these limitations and is of greater utility to the pharmaceutical scientist.

Scanning Electron Microscopy

The scanning electron microscope (SEM) provides a very versatile and powerful imaging ability. Resolutions on the order of 2 to 5 nm, or even smaller, have been obtained by commercial instruments (11). In addition to their high resolution, an important feature of SEM images is the three-dimensional appearance of the images. This comes about due to the large depth of field achieved by the instrument. This depth of field may be one of the most valuable features of this method of microscopy.

The typical commercial SEM consists of a lens system, electron gun, electron collector, cathode ray tubes (CRTs) for displaying the image, and electronics for digital storage of the image. The SEM operates by scanning a fine beam of electrons of medium energy (5–50 keV) in parallel tracks across the sample. The collision of electrons with the sample material results in the generation of secondary electrons, Auger electrons, backscattered electrons, light or cathodoluminescence, and X-ray emission. Samples of up to 25 × 25 mm can be imaged at magnifications from 20× up to 100,000×.

The ultimate resolution of the SEM depends on the electron beam diameter, and the intensity of emitted signals depends on the beam current. These two parameters operate in opposition to each other because a small beam can carry a smaller current than a larger beam. Therefore, the operator must balance these controls to produce the desired image result. The electron beam produced by the electron gun is too large to produce a sharp image. Electron lenses are used to focus the beam into a desirable spot size. The SEM chamber is operated under vacuum to minimize scattering of the electron beam by gas molecules in the chamber. Evacuation of the sample chamber necessitates special preparation of the sample.

The electron beam is scanned across the sample with two pairs of scan coils that deflect the beam. The sample signals most often used are secondary electrons and backscattered electrons. These electrons are collected by an Everhart–Thornley (E–T) electron detector. The E–T detector is electrically isolated from the chamber and has a positively charged wire mesh screen at its entrance. The positive charge attracts the low-energy secondary electrons to the detector. Once they enter the detector, the electrons are accelerated and strike a scintillator material, which produces light that is amplified by a photomultiplier tube.

Preparation of organic materials for SEM imaging requires a trade-off between instrument performance and sample integrity. Organic materials are typically hydrated and composed of elements with low atomic numbers. They may have low thermal and electrical conductivity. Thus, these samples are prone to damage when confronted with a high-energy electron beam. The samples must be dried and coated with conductive material if they are not to compromise the operation of the SEM. An environmental SEM at higher pressures, as discussed in the next section, can be used when sample treatment is not possible. Samples can be dried by chemical dehydration, freeze-drying, critical point drying, or ambient-temperature sublimation. The samples must then be coated with a conducting material to reduce electric charge buildup on the samples. Charging of the sample may cause dielectric breakdown of the sample, thus altering its morphology. Charged areas of a sample may degrade the image resolution by deflection of low-energy secondary electrons or deflection of the electron beam. Samples are typically made conductive by sputter coating with a thin layer of gold or palladium. Heating of the sample from the low-energy electron beam is not typically an issue but could alter heat-labile samples.

Environmental Scanning Electron Microscopy

The SEM offers many advantages, as discussed in the previous section. A drawback to the conventional SEM, however, is that a significant amount of sample preparation is required prior to imaging. In many cases, sample preparation may simply be an inconvenience, but in some cases, the freezing, drying, and coating processes may alter the surface morphology of the sample. The environmental SEM (ESEM) provides the ability to image samples in their wet state, eliminating many of the SEM preparation steps.

Environmental SEMs operate at pressures ranging from 0.1 to 1 kPa. The electron column remains under vacuum through the use of pressure-limiting apertures within the lens (12). Some samples can be imaged directly by using the ESEM, whereas others do require some care in the performance in the instrument. Sample hydration in the ESEM chamber must be monitored as the chamber is prepared. The ESEM chamber can be flooded with saturated vapor during pump-down to ensure continued hydration of the sample. A number of pharmaceutical specimens have been successfully imaged by the ESEM, including biodegradable polymeric matrices (polyanhydrides and lactide-co-glycolide copolymers), hydrogels, and pellet-based formulations (13). In the case of pellet formulations, the ESEM has allowed imaging of hydrated film coatings of pharmaceutical pellets. In contrast, the conventional SEM would not allow imaging of coating films in their hydrated state.

Scanning Tunneling Microscopy and Atomic Force Microscopy

The scanning tunneling microscope (STM) and the atomic force microscope (AFM) operate on the same quantum mechanical principle. The 1986 Nobel Prize in Physics was awarded to Gert Binnig and Heinrich Rohrer for the development of the STM as well as to Ernst Ruska for work on the electron microscope. In the STM, a pointed metal probe is placed within a few nanometers of the sample. With a small (10 mV) voltage applied to the probe, electron tunneling can occur between the probe and the sample. An exponential relationship exists between the separation between the probe and sample and the tunneling probability such that changes in separation as small as 0.01 nm can result in measurable changes in tunneling current. By scanning the probe over the sample surface, bumps on the surface can be measured and a topographic map can be generated. Piezoelectric ceramic transducers are used to perform the scan because of their precise ability to position the probe. The STM requires that the sample be conductive (14).

A variation of the STM, the atomic force microscope can be used with insulators. The AFM has been used to characterize the surface roughness of pharmaceutical particles (15) and the elastic properties of aerogels (16). In an interesting application of the AFM, probe tips were mounted with spherical colloidal silica particles, and the forces of interaction between the colloid probe and other pharmaceutical materials were measured (17).

Indirect Imaging

Indirect imaging can be used to measure the actions of a single particle at a time or to collect data from all particles in a sample at the same time.

Single Particles

Single-particle methods of indirect imaging consist of means of measuring the actions of a single particle at a time. Individual measurements of multiple particles clearly must be taken to develop a distribution. Single-particle techniques do not measure the entire powder or aerosol. Thus, considerations must be made concerning the validity of sampling. Techniques that use single-particle methods include electrical resistance, light blockage, and laser velocimetry.

Electrical Resistance

Electrical resistance (electrozone) particle-sizing methods are used with particles suspended in an electrolyte solution. The Coulter counter, a well-known version of this type of instrument, was developed originally for blood cell counts. In its operation, particles in the electrolyte solution are forced to flow through a small orifice. Electrodes placed on either side of the orifice maintain constant current flow through the orifice. The passage of a particle through the orifice causes the resistivity to increase as the particle displaces the electrolyte. This resistivity increase results in a voltage spike in which the amplitude of the spike is proportional to the volume of the particle. Pulses are classified by area, and a count is incremented in a bin corresponding to the size. Counting of multiple individual particles results in a distribution.

The electrozone technique has a dynamic range of sizes from 0.5 to 800 μm, but different orifice sizes must be used to achieve this range. A given orifice can be used with particles ranging in size from around 2 to 50 percent of the orifice size; blockage of the orifice becomes a problem with larger particles, and smaller particles are undetectable above electronic noise. The resolution decreases near the lower limit of the measurable range. The principle of operation is based on the displacement of fluid, so the particle volume will be measured regardless of shape. The instrument is easily calibrated with polystyrene size standards. If two particles pass through the orifice at the same time, however, they will be counted as a single particle (18). Other drawbacks include the requirements that samples must be suspended in conductive fluids and must be insulating materials.

Light Blockage

The light-blockage technique of particle sizing is the optical equivalent of the electrical resistance method. In a light-blockage instrument, particles in the liquid are passed through a counting chamber, where they partially block a laser beam focused on the chamber. The reduction in light intensity measured by a detector is proportional to the optical cross-section of the particle. Like the electrozone counter, a population of single particles is measured.

This technique is applicable to a broad range of particle sizes spanning from 0.5 to about 2,000 μm, although different size sensors must be used. Analysis times are rapid, on the order of 3 minutes or less. Resolution is decreased for nonspherical particles because the particle cross-section is evaluated. For nonspherical particles, the orientation of the particle as it passes through the detector will also have an effect on the measured size.

Laser Velocimetry or Time-of-Flight Counter

The time-of-flight counter, or laser velocimetry method, determines particle size by measuring the transit times of individual particles between the beams of two lasers. The technique is used most often with dry powders, although it may be used with nebulized suspensions. Laser velocimeters operate by accelerating particles through a jet or nozzle into a partial vacuum. The particles accelerate according to their size, with smaller particles accelerating faster than larger particles. When a particle passes through the first laser beam, a time-of-flight clock is started and then stopped upon the particle's passage through a second beam. Particle size is calculated from the measured time-of-flight.

The time-of-flight technique is capable of measuring particle sizes ranging from about 0.2 to 700 μm, although resolution is limited at the bottom of the range. The time-of-flight counter can be calibrated readily with solid particle size standards. Measurement times are rapid, normally about a minute. Some instruments are capable of measuring on the order of 100,000 particles per second (19). Care must be taken to introduce the appropriate concentration of particles because the counting speed of the instrument is limited (20). Nonspherical particles will not be sized correctly, but the magnitude of the error is difficult to determine because it depends on shape factors (21). Due to the high shear forces that are generated, particles are deaggregated by the instrument (22). Therefore, sizing results may not be relevant to powders that tend to aggregate. In addition, the particle density must be known for use in the particle size calculation.

Populations of Particles

Ensemble methods of particle sizing collect data from all particles in a sample at the same time and then calculate size distributions from the data. Most of these instruments rely on the Mie theory of light scattering in their sizing algorithms. Ensemble particle-sizing techniques include light scattering, laser diffraction, ultrasound scattering, and gas adsorption.

Light Scattering

Light scattering, or photon correlation spectroscopy (PCS), can be used for ensemble sizing of small particles. PCS operates on the principle that light scattered from small particles varies in intensity due to the Brownian motion of the particles. The technique is applicable only to particles for which movement due to Brownian motion is more significant than movement due to sedimentation, such as particles smaller than 3 μm. In this method, particles are dispersed in a liquid and placed in a sample holder that is illuminated by a laser beam. (23) A photomultiplier tube mounted off-axis from the laser measures scattering intensity. Brownian motion of the particles causes fluctuations in scattering intensity. This intensity is autocorrelated over short time scales to obtain the diffusion coefficient for the distribution, which is inversely proportional to the decay time of light scattering. The diffusion coefficient, D, is related to the particle diameter, d, by the Stokes–Einstein equation:

$$d = \frac{k_B T}{3\pi\eta D} \tag{5.2}$$

where k_B is the Boltzmann constant, T is temperature, and η is the fluid viscosity.

PCS can be used with very small samples of material. Due to the short time scales of Brownian motion, analysis is very rapid. The temperature of the sample must be precisely monitored because the diameter calculation depends on T.

Laser Diffraction

Most laser diffraction particle sizing is based on Fraunhofer diffraction, a special case of the Mie theory (24), although some instruments use Mie theory. Fraunhofer diffraction calculations provide the volume of particles given their diffraction pattern. In this sizing technique, laser diffraction instruments pass a laser beam through a sample dispersed in liquid or air (25). The focused laser beam is diffracted by particles in the beam, more broadly by smaller particles and less so by larger particles. The scattered light is focused by a lens onto a ring array diode detector that passes scattering intensity data to a computer for particle size distribution determination.

Early laser diffraction instruments used a series of lenses to measure ranges of particle sizes. Instruments currently available can measure a very broad range of particle sizes. In either case, laser diffraction can measure a dynamic range of particle sizes from 0.5 to 1800 μm. Most laser diffraction instruments have accessories for measuring dry powders dispersed as aerosols and nebulized solutions or suspensions. They can be arranged for synchronization of the metered dose inhaler spray and the data collection. Alternatively, the particles may be suspended in a liquid for measurement.

Laser diffraction instruments have been widely used, primarily because of the ease of measurement. Data collection with this technique is very rapid. Data are expressed in terms of volume distributions, although other parameters may be calculated. Errors due to multiple scattering may occur if the concentration of particles is too large. Multiple scattering is the effect by which light scattered from one particle is scattered by a second particle before reaching the detectors. The particle concentration in the laser beam must be chosen to provide sufficient diffracted light but not be high enough to cause significant multiple scattering events. The index of refraction of the particle must differ from the suspending medium (air or liquid). Latex spheres of stated size guaranteed by the manufacturer may be purchased for calibration. It is also possible to purchase a reference reticle from the instrument manufacturer. This reticle has a polydisperse particle size distribution of chrome dots on quartz with a diffraction pattern determined by the manufacturer from a properly calibrated and aligned instrument (25).

Particle size distributions from laser diffraction instruments are based on calculations from theory. All instruments base this calculation on Fraunhofer or Mie diffraction theory, but the precise algorithm is typically considered proprietary and not provided to the user. This leads to the tendency to treat these instruments as "black boxes." Another technique, such as microscopy, should always be used to confirm light scattering results.

Ultrasound Sizing

Ultrasonic energy can be used to perform size and chemical characterization of particles in liquid (26). Acoustic spectrometry calculates particle sizes from scattering of ultrasound energy in exactly the same manner as particle sizing is determined by using light scattering. A particular advantage of ultrasound characterization is the ability to measure particle size in liquids, including opaque dispersions. Acoustic spectroscopy has a dynamic range of 10 nm to 1 mm. As such, the technique can be used for in-line process monitoring. A drawback to acoustic spectroscopy is the long acquisition times (up to 3 minutes) required to take a full sample measurement.

Gas Adsorption

Gas adsorption studies can be used to determine the surface area of an adsorbing solid. Interest in porous particles for aerosol delivery makes gas adsorption an especially important technique (27). If it is assumed that particles are spherical, the volume-surface diameter can be calculated from the surface area (28). If ρ is the density of the adsorbent material, and S_w is its calculated specific surface area in square centimeters per gram, then the volume diameter, d_v, is given by:

$$d_v = \frac{6}{\rho S_w} \tag{5.3}$$

Physical Methods of Particle Size Separation

Physical methods of particle size separation take advantage of fundamental physical properties of particle dynamics to segregate particles by size. Most of the methods described in this section utilize the settling properties of particles in a gravitational or other force gradient field. Some of the techniques discussed offer the potential to collect size-segregated particles for further use. Others, such as the impaction methods, provide information relevant to particular states of the powder, that is, the aerodynamic properties of the powder dispersed as an aerosol.

Sieving

Sieving is a method of particle size analysis that can also provide size-segregated particles for use in processing. The method is described in Chapter 4, along with potential measurement errors. Results are presented at cumulative percent undersize for the aperture diameter of the given stage (2).

Sieves are supplied according to a standard, such as the British Sieve or the International Organization for Standardization (ISO) standard. The US standard is a $\sqrt{2}$ progression of sizes starting at 350 μm down to 45 μm. Sieves are also classified by their mesh size, measured by the number of wires per inch. Sieving results may vary as a function of sieve load, particle shape, agitation method, physical properties, and overall size distribution. For example, Hickey and Concessio have noted that some pharmaceutical powders tend to aggregate when sieved by vibration versus tapping (29). Other errors may occur due to errors in sieve construction or due to wear. In general, if losses in sieving exceed 0.5 percent of the original sample weight, the data should be discarded.

Inertial Impaction

Inertial impaction is most frequently used in pharmaceutical science to size particles intended for aerosol dispersion. It is also used extensively in environmental science to sample naturally occurring or human-produced aerosols such as pollutants. Inertial impaction utilizes particle inertia to segregate aerosol particles into discrete size ranges (30). Impactors were initially developed in the 1940s to analyze environmental aerosols (31). These devices consist of circular orifices of decreasing diameter arranged on stages in series. An impaction plate is positioned beneath each orifice stage. The airflow velocity through the orifices increases as their diameter decreases because the system is closed and the air volume flow rate drawn through the impactor is constant. If aerosol particles are to remain entrained in the airflow through the device, they must follow the curving air streamline through the orifice and around the impaction plate. Particles with an inertia too great to achieve the angular acceleration required to remain in the streamline will continue in their initial direction and impact on the collection plate. Particles with a smaller aerodynamic mass, and thus a smaller inertia, will continue to the next stage of the impactor. The process continues through the impactor, and the particles are segregated by size and inertia as the airflow velocity increases through the narrowing jets.

The particle size measured by inertial impaction is the aerodynamic size. The aerodynamic size of a particle is a function of the balance of aerodynamic and gravitational forces on a particle, given its size and density. The aerodynamic diameter of a given particle is defined as that of a unit density sphere with an equivalent settling velocity, such that:

$$\rho_a d_a^2 = \rho d^2 \tag{5.4}$$

where ρ_a and ρ are the aerodynamic and true densities of the particle, respectively, and d_a and d are the aerodynamic and geometric diameters of the particle, respectively. Since $\rho_a = 1\ \text{g/cm}^3$, the equation relating the geometric diameter to the aerodynamic diameter for spherical particles is as follows:

$$d_a = \sqrt{\rho}\, d \tag{5.5}$$

For nonspherical particles, a shape factor must be included in equations (6.4) and (6.5). A shape factor ranging from approximately 1.1 to 1.75 is frequently used as a multiplicative term on the right-hand side of equation 6.5. The practical and theoretical implications of shape on aerodynamic properties have been discussed by many authors (32–34).

Inertial impaction is the particle sizing method prescribed by compendial specifications for medical aerosols (35–37). Its utility stems from three major factors. First, the entire aerosol generated by an inhalation device can be sampled by inertial impaction. Second, the collection of particles in the impactor is assayed by chemical means so the size distribution can be calculated for drug particles only. Finally, the size measured by impactors is the aerodynamic size. The aerodynamic size is the most relevant to inhalation because this is the size that influences deposition in the lungs.

When sizing aerosols to be generated by inhalation devices, care must be taken that the aerosol is representative of the aerosol that would be generated by the device in clinical use. This specification may conflict with the very specific operational parameters under which impactors are calibrated. For example, the Andersen cascade impactor (ACI) is calibrated at flow rates of 28.3, 60, and 90 liters per minute (Lpm); a flow rate of 28.3 Lpm may not be sufficient for dispersion from a dry powder inhaler (DPI), whereas 60 Lpm is considered achievable by most patients (38). Another drawback to inertial sizing is that the collection of the particles must be performed manually. The procedure is very labor intensive and time consuming. However, it should be noted that the Next Generation Impactor (NGI), which samples at similar flow rates to the ACI (see below), collects the aerosol fractions in dishes that can be adapted for greater automation and speed of sampling and analysis. Finally, impactors distribute an aerosol into discrete size ranges, typically ranging from 0.2 to 10 μm. A larger range of sizes or finer resolution might be preferable for some applications, such as sizing of nasal sprays.

Of the various impactor types available, the examples provided in the USP include the Andersen Mark II nonviable 1 ACFM ambient sampler (ACI) and the NGI (36, 37). These impactors have eight and seven stages, respectively, and are calibrated at fixed flow rates. At 28.3L/min (1ACFM), particles are separated by the ACI into fractions with cut-off diameters (d_{50}) of 9.0, 5.8, 4.7, 3.3, 2.1, 1.1, 0.7, and 0.3 μm. Particles that do not impact on the last stage are collected in a backup glass fiber filter with a pore size of 0.2 μm. At 30L/min, the NGI separates particles into fractions by cutoff diameters of 11.7, 6.4, 4.0, 2.3, 1.4, 1.0, and an exit micro-orifice collector (MOC). The ACI and NGI are also available with modified initial stages calibrated for use at 60 and 90 Lpm. At 60 Lpm, the cutoff diameters for the ACI stages are 8.6, 6.2, 4.0, 3.2, 2.3, 1.4, 0.8, and 0.5 μm and for the NGI are 8.0, 4.5, 2.8, 1.7, 0.9, 0.5, and 0.3 μm (MOC, 0.1 μm). When operated to collect dry powder aerosols, both impactors are equipped with a preseparator stage that has a cutoff diameter of 9.9 μm for the ACI and 11.7 (30L/min) and 8.1 (60L/min), respectively, for the NGI. The preseparator is designed to remove very large particles from the aerosol cloud. These particles would be subject to bounce and reentrainment and could overwhelm the first stage of the impactor. Many authors have evaluated the Andersen impactor for calibration accuracy, manufacturing errors, and operational errors (39–41). Notably, the ACI was adopted for use as an environmental sampler whereas the NGI was the first pharmaceutical aerosol sampler. The NGI has the advantage of the collection surface being a dish into which known quantities of medium can be introduced to allow efficient collection and accurate estimates of deposited masses of drug or other material. The ACI employs flat plates as collection surfaces and requires a separate step, usually petri dishes or sealed bags, into which both the plate and the medium can be introduced. Under adequately controlled conditions with well-trained staff, this can be an efficient method, but the additional step could lead to either operator or sampling error.

Errors in particle sizing using these impactors, and most other impactors, result from particle bounce and wall losses (39, 42). Particle bounce occurs when particles, especially dry particles, are not captured by the collection plate with which they collide, but are instead re-entrained in the airstream and collected at a lower stage. Particle bounce can be minimized by coating the impaction plates with silicon oil (43). Wall losses (interstage losses) in impactors are a size-dependent phenomenon wherein particles impact the walls of the impactor as they transit from one stage to the next or before moving through a jet (42). Wall losses cannot be eliminated. The USP specifies that if interstage losses account for more than 5 percent of the total dose from a device, the interstage mass can be included with the associated impaction plate in size distribution calculations.

Another type of inertial impaction instrument is the liquid impinger. Impingers use a liquid as the inertial impaction surface. As such, they are well suited for use with powder aerosols because impinging particles are captured by the liquid, thereby eliminating the possibility of particle bounce. In addition, sample collection from liquid impingers is less tedious than for impactors; the liquid is removed with a pipette and analyzed. Commercially available impingers have fewer stages than impactors so resolution of particle size is more limited. A four-stage liquid impinger is one example (44). The two-stage liquid impinger is convenient for rapid analysis of the fine particle fraction or "respirable dose" of an aerosol (45). The fine particle fraction is defined differently by various sources, but in general, it is the fraction of particles in an aerosol of a size small enough to penetrate the lungs. The two-stage liquid impinger separates particles into two size ranges, those greater than and those less than 6.4 μm (46).

Filtration

Filtration may be used either for particle size determination or for particle segregation for other purposes (47). The use of a graded sequence of filters in cascade provides particle sizing data much like sieving. The pressure drop at each filter stage must be monitored to ensure that clogging does not occur anywhere in the cascade. Filters are also useful for the collection of aerosol particles for microscopy. Membrane filters are particularly useful for microscopy because their structure, consisting of tortuous channels, tends to collect particles on their upper surface (1). The filter can then be used directly for microscopy. Filters operate on the principle of interception and impaction. Large particles impact directly on filter fibers. Smaller particles may avoid initial impact but are intercepted as they leave the streamlines of air passing around the filter fibers.

Andreasen Pipette

The Andreasen pipette instrument uses a sedimentation technique for particle sizing. The settling velocity of particles in a fluid depends upon their diameter. The Andreasen pipette takes advantage of this effect by measuring concentration changes at the bottom of a vessel as suspended particles settle in the dispersing liquid. The sedimentation vessel is filled with a powder sample dispersed in liquid to a height of 20 cm. In order to avoid aggregation effects in the liquid, the sample concentration should be 2 percent or less by volume (28). The powder sample is dispersed by shaking or, preferably, by slowly turning the vessel upside down and back for a number of repetitions. Sampling should not begin for one minute, to allow particles that may have had an initial velocity to reach their settling terminal velocity. A 10 mL sample is then drawn from the pipette, generally at 20-second intervals. The sample is then drained, dried, and weighed or otherwise analyzed for concentration. Particles collected by the tube at height (h) are between the sizes that settle in the time between t_i and t_{i+1}, where the times are for the i_{th} and $(i + 1)_{th}$ samples. This particle diameter, d_p, is derived from Stokes' equation and is as follows:

$$\sqrt{\frac{18h\eta}{(\rho_p - \rho_f)g^{t_{i+1}}}} \le d_p \le \sqrt{\frac{18h\eta}{(\rho_p - \rho_f)gt_i}} \tag{5.6}$$

where η is the kinematic viscosity, g is the gravitational constant, and ρ_p and ρ_f are the densities of the particle and the fluid, respectively.

The Andreasen pipette method is inexpensive. However, it does have drawbacks: Potential for error occurs because the sampling tube interferes with sedimentation, especially directly below the tube, and because the tube samples a spherical rather than a planar region. The sedimentation fluid must be kept at a constant temperature for the duration of the experiment because viscosity is a function of temperature. In spite of these limitations, the method can still be accurate to within 2–5 percent.

Virtual Impinger

The virtual impinger was developed to minimize the bounce and re-entrainment problems of the cascade impactor (2). In this impinger, no collection surfaces are present. Instead, an airstream is directed toward a container with a volume of static air. The airstream is diverted around the sides of the vessel because air cannot be conducted through it. This static air forms a virtual impaction surface. Larger particles will be carried by their inertia out of the streamline and into the vessel, where they are collected. Virtual impingers operate with less turbulence at the vessel edges if there is a slow flow of air from the vessel. One example is the dichotomous sampler in which 1/49th of the total flow passes through the collection vessel. A collection filter is placed in the vessel to capture particles entering the opening. Virtual impingers can be placed in series with an increasing airflow rate moving down the series to produce a particle size distribution measurement.

Centrifugation

Centrifugation works on the same principles as sedimentation, except that the gravitational force is replaced by a centrifugal force generated by a spinning sedimentation instrument. Forces much greater than gravity can be generated with the centrifuge to increase settling and reduce experimental time. As a consequence, small particles that would settle slowly under the influence of gravity can be induced to sediment more quickly under a much higher centrifugal force. The gravitational term in Stokes' equation [see equation (5.6)] for particle terminal velocity is replaced by a centrifugal term, a_r, as follows:

$$v = \frac{2a_r d^2 (\rho_p - \rho_f)}{9\eta} = \frac{2\omega^2 r d^2 (\rho_p - \rho_f)}{9\eta} \tag{5.7}$$

where ω is the radial velocity and r is the radial position of the particle relative to the center of rotation. Since particle velocity is a function of radial position as v = dr/dt, equation (5.7) can be rearranged to yield:

$$\frac{dr}{r} = \frac{2\omega^2 d^2 (\rho_p - \rho_f)}{9\eta} dt \tag{5.8}$$

Integrating equation (5.8) and solving for particle diameter d yields the diameter of a particle with initial position r_1 and position r_2 at a time t:

$$d = \sqrt{\frac{9\eta \ln(r_2 / r_1)}{2\omega^2 (\rho_d - \rho_f) t}} \tag{5.9}$$

This diameter is used as the diameter of particles sampled from the centrifuge at time t.

In operation, the centrifuge is brought to the chosen speed, and the spin fluid is introduced from a coaxial injection needle. Once the fluid has stabilized, the powder sample dispersed in the spin fluid is injected into the centrifuge. Samples are taken at time intervals, concentrations measured, and diameters calculated from equation (6.9). An advantage of this approach over gravitational sedimentation is the accuracy with which a sample can be introduced. The sample is introduced with zero initial velocity and without turbulence. The density and viscosity of the spin fluid can be chosen to achieve reasonable centrifuging times. Centrifugation gives results accurate to less than 2 percent uncertainty in the cumulative weight undersize distribution. As with gravitational sedimentation, the sample concentration should be less than 2 percent to avoid hindrance of sedimentation. The particle density must be known in order to calculate diameters. Errors in its measurement will contribute significantly to errors in the results. Ultracentrifuges that spin at speeds of up to 60,000 revolutions per minute (rpm) can be used to measure particle size distributions for particles as small as 0.01 μm.

Approaches to Calibration

Closely related to the measurement of particle size, distribution, and morphology is the need to calibrate the instruments performing these tasks. For direct imaging techniques, calibration is straightforward. A particle of known diameter is required, or precisely manufactured graticules or micrometer stages may suffice. Microscopy calibration standards include manufactured spheres and naturally occurring particles. Manufactured standards are constructed from glass, quartz, and polystyrene. These standards may be purchased in sizes ranging from 0.4 to hundreds of μm. Pollen grains are nearly monodisperse and can be used for calibration. Pollen grains range in size from approximately 15–80 μm.

Calibration techniques for other methods of particle sizing are more complex and may require specialized equipment. For these instruments, the physical and chemical properties of standards may need to be precisely controlled. Particle generation for calibration of aerosol particle sizing equipment requires continuous generation of particles over the entire sampling time. They must further provide an aerosol of sufficient concentration for measurement by the instrument.

Vibrating Orifice Monodisperse Aerosol Generator

The vibrating orifice monodisperse aerosol generator (VOAG) creates particles via the breakup of a jet of liquid forced through an orifice that is driven by a periodic vibration. Vibration of the orifice is provided by a piezoelectric ceramic. This vibration breaks up the jet into uniform droplets. The resulting stream of particles is diluted with heated air to evaporate the solvent and minimize droplet coagulation.

The liquid feed rate, Q, and the frequency of vibration, f, are the primary determinants of the generated particle diameter, given by:

$$d = \left(\frac{cQ}{\pi f \rho_p} \right)^{\frac{1}{3}} \tag{5.10}$$

where c is the mass concentration by volume of solute and ρ_p is the density of the particle. Particle diameters ranging from 0.5 to 50 μm can be achieved (48). Particle size distributions are monodisperse with $\sigma_g < 1.05$. Larger particle production is not practical because evaporation time is lengthy for the production of solid particles and concentrations are low. Typical particle concentrations generated with a 10 μm orifice are around 10^3 cm^{-3}. Concentration decreases with decreasing orifice size because the vibration frequency also decreases (and, thus, the number of particles generated per second decreases).

The liquid fed through the orifice should be filtered. However, orifice plugging can still be a problem. For solid particles where evaporation is required, appropriate choices in solvent, concentration, and compound are important. Evaporation rates that are too rapid will result in the production of less solid and rough-surfaced particles.

Spinning Disk Generators

The spinning disk method of monodisperse particle generation produces less concentrated aerosols (around 100 cm^3) than the VOAG instrument. It does have the advantage that, without an orifice, plugging is not an issue, so operation is very reliable. The spinning disk is a rotor–stator system in which high-pressure air is directed over vanes on the rotor, causing it to spin rapidly, up to 70,000 rpm. A liquid feed is directed onto the top center of the disk, so droplets are flung outward by centrifugal force. Typically, larger primary droplets and smaller satellite droplets are generated. The satellite droplets are deflected by the airstream and rejected.

Droplet sizes ranging from 10 to 150 μm are typical. The size is a function of the rotation angular velocity (ω), liquid concentration (c), rotor diameter (D), and liquid surface tension (T), according to the equation:

$$d = \frac{K}{\omega}\left(\frac{T}{D\rho_l}\right)^{\frac{1}{2}}\left(\frac{c}{\rho_p}\right)^{\frac{1}{3}} \quad (5.11)$$

where the constant (K) is an empirical term ranging from 2.3 to 7 for water (49). The densities (ρ_l and ρ_p) refer to the liquid and the particle, respectively.

As stated, suspensions can be used with the spinning disk more readily than with the VOAG because the disk will not become plugged. However, the cautionary statements concerning particle morphology as a function of evaporation rate apply to this instrument as well.

Charge Neutralization

The generation of monodisperse particles by charge neutralization relies on the balance of electrostatic and aerodynamic forces on particles. In operation, a polydisperse aerosol is passed through a charging chamber containing a radioactive source (usually Kr^{85}). Particles receive a single positive, negative, or zero charge. The particles are then passed to a differential mobility analyzer that has a grounded sheath surrounding a central, negatively charged rod. Negatively charged particles are attracted to the outer sheath, positively charged particles are attracted to the central rod, and uncharged particles pass with the airflow and are pumped out of the chamber. Particles of an appropriate diameter will move toward the central rod but will just miss the rod's end and instead be pulled into a tube below the rod. By adjusting the flow rate and voltage of the central rod, the aerodynamic and electrostatic forces on particles can be balanced to "dial in" the desired monodisperse particle size output.

PARTICLE SIZE DIMENSIONS AND STATISTICS

Particle size analysis has been the subject of a number of excellent texts (50–52). Particle size clearly plays a role in formulation development (53–55).

Preparation of particles for use in pharmaceutical products involves a number of different processes. A general scheme may be outlined as shown in Figure 5.3.

The physicochemical properties of the drug or excipient play a role in the adoption of conditions under which each of these processes is conducted (56, 57). The initial solution or suspension characteristics are dictated by the solubility and dissolution of the drug in the selected solvent, which in turn relates to the polymorphic form of the starting material (58–60). Nucleation is influenced by the presence of impurities and lattice formation. If the starting particle size is not suitable for preparing a pharmaceutical product, size reduction is required. Milling involves mechanical energy input to reduce the size of particles (47, 61, 62). Fractures in the crystal lattice create new surfaces, and recrystallization phenomena thus may occur during the manufacturing process (63–65). Specifications on the particle size required for a product involve a defined size range, which may require active segregation (sedimentation, sieving, electrostatic precipitation) (52). This requires either overcoming the cohesive nature of the particles or the formation of stable aggregates (66–68). Once the particles of each of the components (drug and excipients) have been produced, they are blended to uniform drug dispersion (69–72). Segregation and blending methods may result in some attrition of particles, and changes in particle size may occur.

Particle size plays an important role in each of these processes. In addition, stable aggregate size may play a role in the performance of the product. In this case, the

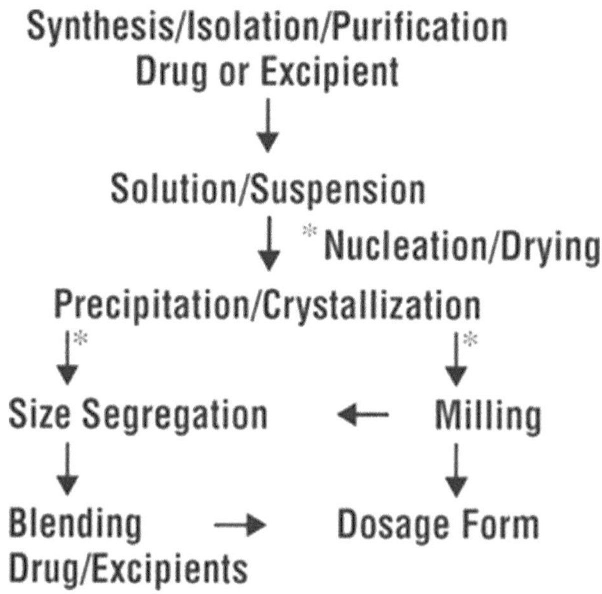

FIGURE 5.3 Flow diagram for common processes involved in the preparation of particles suitable for inclusion in pharmaceutical products.

*Indicates points in the processes where nucleation/drying occur.

forces of interaction between particles must be considered (73). These include ionic, covalent, charge/complexation, van der Waals, capillary, and mechanical interlocking forces (74). Particle interaction influences aggregation flow (68), dispersion (66, 67, 75), compression (72), moisture content, sampling (76), and statistical descriptors of powders.

Ultimately, the distribution of drug particles in a blend impacts the quality criteria for the unit dose (77). The amount of drug a patient receives in the dosage form (e.g., tablets and capsules) should reflect the label claim for that product. Official specifications cannot simply be the subject of regulations independent of science and technology. Quality control can be maintained only with a basic understanding of the pharmaceutical technology involved in manufacturing.

Variations in the amount of drug that a patient receives from a single dose depend on two factors, actual volume or weight of therapeutic agent and degree of homogeneity obtained as a result of the mixing processes. Homogeneity in this context is something of a misnomer. The homogeneity of a blend or suspension, unlike the molecular dispersion of a solution, is limited by the degree of subdivision of the drug particles. The distribution of drug particles within the diluent dictates uniformity. Particle behavior, also a function of particle size, may be a factor in uniformity. A suspension, for example, may be more or less stable based on particle size and density with respect to the dispersion medium. The ability to redisperse the drug particles following storage may also be influenced by their characteristics.

The influence of particle size and distribution on pharmaceutical processes and ultimately on the efficacy of drugs delivered from solid dosage forms cannot be discussed without considering the methods for describing these parameters. Particles of any substance exhibit unique morphologies. These morphologies are usually related to the chemical composition of the particles and the method of manufacture and processing (47). This chapter will not consider the ways in which different particles are obtained, which was addressed in earlier chapters; rather, it will focus on the way the particle morphology contributes to estimates of particle size.

Real particle populations consist of heterogeneous mixtures. Individually classifying particles by size, from smallest to largest, may result in a shift in shape, surface rugosity, and energetics. Since these characteristics may play important roles in particle performance, we may infer that a good understanding of particle size and distribution is important to process development.

Some methods of manufacture, such as spray drying, result in spherical particles (56, 78, 79). Particles with spherical morphology and smooth, nonporous surfaces are easily characterized because their diameters unambiguously describe their sizes (50). This ease of characterization derives from the infinite rotational symmetry associated with these particles. Simply stated, it does not matter how you measure the size of these particles or who conducts the study. Within the limits of error of the scale of measurement, the same answer will be obtained under all circumstances. For the moment, this ignores the issue of sampling (76). As noted in earlier sections, however, most pharmaceutically relevant solids are crystalline in nature and exhibit a range of crystal systems and habits, resulting in a variety of shapes and surface irregularities (80).

PARTICLE SIZE DESCRIPTORS

In order to communicate clearly the particle size and distribution of irregularly shaped particles, a "language" of descriptors must be agreed upon by those working in the field. This raises the question, "How many different ways can we consider the dimensions of a particle?" The best way to address this question is to take into account the intended purpose of the particles.

PARTICLE SHAPE

Irregular particles are difficult to reproduce accurately in terms of their surface invaginations. Sophisticated attempts to describe particle peripheries have been attempted by using polar coordinates from which Fourier coefficients may be derived or by using a fractal approach (81, 82). However, these methods are tedious and lack broad application. Other attempts at irregular particle shape description involve Heywood and Waddell shape factors, which are described in Chapter 6 (83).

PARTICLE SIZE

The convention for describing particle size has been to consider irregular particles in terms of spherically equivalent particles with the same surface area, volume, and projected area, which are static measures, or with the same sedimentation rate and aerodynamic behavior, which are dynamic measures. The reason for considering each of these equivalent parameters relates to the application of the data. For example, surface area may be important for catalysis, whereas sedimentation may be important for the stability of a suspension formulation.

The particle size descriptor for individual particle sizes can be traced back to the method by which it was determined. The following brief review of some key particle size descriptors is discussed in more detail in Chapter 6.

Microscopic analysis of particles has resulted in the determination of a variety of particle size dimensions, including Feret's diameter, Martin's statistical diameter, and Heywood's projected area diameter (84, 85). Feret's diameter may be defined as the distance between two tangents drawn at the periphery of the particle perpendicular to the direction of the scan. Invoking the direction of the scan was anticipated to allow accurate mean data to be derived because variations in the shape of particles would be averaged by their random orientation. However, this dimension frequently overestimates the population mean size. Martin's statistical diameter involves bisecting a particle by area in the plane of scan and then measuring the length of the line as it crosses the particle periphery. Ultimately, the measure that found most favor for measuring the geometric size of particles from visual observation was Heywood's projected area diameter or simply the projected area diameter. This measure involves equating the area of a circular disc to the projected area of the particle and using the diameter of the disc as the characteristic dimension of the particle. Figure 5.4 illustrates the diameters for comparison.

To simplify the estimation of projected area diameter, graticules have been developed for light microscopy, which shows discs in $\sqrt{2}$ scales from which sizes can be

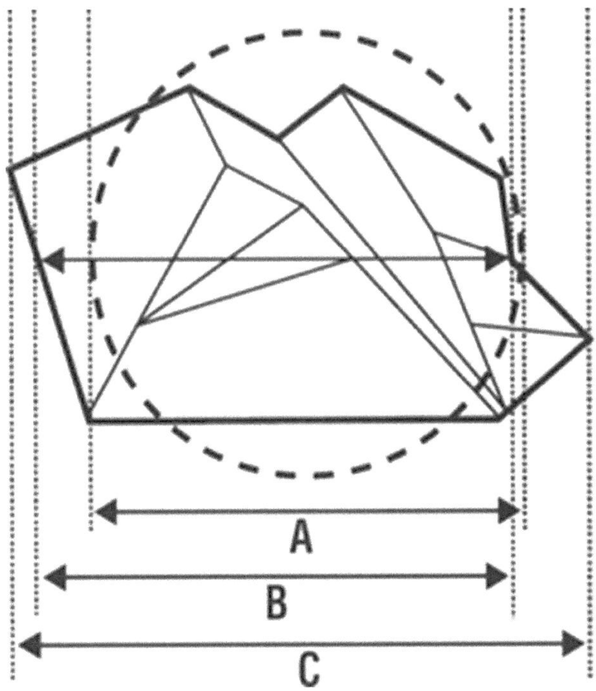

FIGURE 5.4 Characteristic particle diameters: (A) projected area, (B) Martin's statistical, and (C) Feret's.

estimated directly. The most commonly used graticules are the Patterson Globe and Circle or Porton graticules (86). An important factor in this estimation is that particles being examined microscopically are generally in the plane of greatest stability, which means the third dimension in space, depth, is not being assessed. For plates and fibers, this may lead to misleading estimates of particle size.

Sedimentation of particles is known to be a function of Stokes' law in that particles are subjected to gravitational acceleration (g) and resisted by the viscosity (η) of the medium through which they are falling. The general expression of Stokes' law governing this phenomenon is as follows:

$$\frac{h}{t} = \frac{\rho d_{st}^2 g}{18\eta} \tag{5.12}$$

where t is the sedimentation time, h is the height through which the particle falls, d_{st} is the Stokes diameter of the particle, and ρ is the difference in density between the particle and the medium.

Consequently, the Stokes diameter can be derived from the height through which a particle will fall as a function of its size and time of sampling. This is a particularly important parameter for suspension stability and may be used to select bulking agents and to design processes that require suspended particles.

A variety of individual particle sizes are based on other geometric properties of particles, the most notable among these being surface area and volume. The surface area of a sphere is related to the diameter by the term πd^2, and the volume is related by $(\pi d^3)/6$. Methods such as gas adsorption surface area measurement and electrical resistance may be employed to estimate the equivalent surface area and volume diameters, respectively, of particles being studied (87).

Caution should be taken not to convert particle size data based on one set of particle properties to another, which can lead to significant errors. Wherever possible, the particle sizing technique should duplicate the process under evaluation.

Powder Mixing

In the foregoing section, particle size was viewed in terms of evaluating the individual particle size from which a population of particle sizes can ultimately be constructed. The degree of homogeneity of a powder blend is a key area in which particle size and distribution play a role (70). Two major processes for powder mixing—dry mixing and solvent—are shown in Figure 5.5.

Dry mixing requires control of particle size and distribution, enabling prediction of content uniformity and dissolution rate. Flammable solvents are not required in this

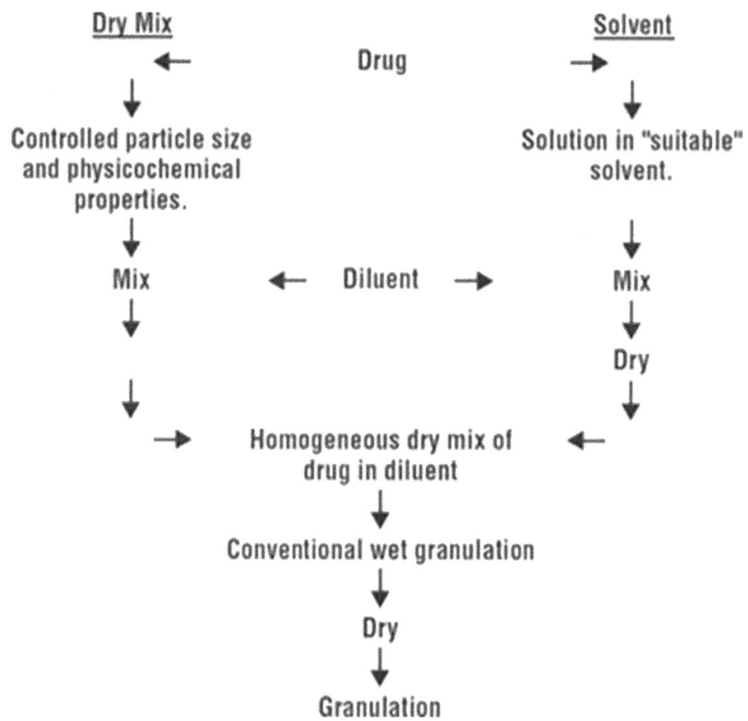

FIGURE 5.5 Principal methods for formulating tablets containing small amounts (<5%) of drug. A suitable solvent is one in which the starting materials are soluble and with a known safety profile. The last step leading to dry granulation would be followed by filling and compression.

method, so the safety hazard is reduced. Homogeneity may be difficult to achieve if fine particles are employed, however, because they may aggregate, causing poor dispersion of the drug. The solvent method theoretically results in a more ordered mixing and dissolution rate. However, the final particle size and physicochemical properties of the product are difficult to predict. Among the possible undesirable outcomes from the application of the solvent method are solvate formation, residual solvent, degradation during processing, and variation in crystal conditions. Solute migration may give rise to drug-rich fractions. In addition, dry mixing is essential subsequent to the solvent method, and flammability may be a hazard during processing. In either case, and for subsequent granulation and compression, a knowledge of particle size and distribution of all the components is important in establishing processing parameters (e.g., mixing; flow and dispersion; the pharmaceutical characteristics, particularly content uniformity and dissolution rate; and specifications on the quality of the final product).

Particle Size Statistics

"Real" or "true" distributions of particles cannot be known due to shortfalls in the experimental approach as well as an inability to capture the entire nature of a particle population. The two general approaches to evaluating particle size are (1) measuring derived properties of individual particles and summing them for a distribution and (2) observing a physical phenomenon that occurs for a group of particles (88, 89). Even when used correctly, these techniques are subject to error. In extreme cases, abuses of the methods can lead to misleading information.

Principles of Statistical Inference

The sequence of events in describing particle size distributions begins with the selection of a particle measurement technique. Once this technique is selected and applied, the sequence of events in the evaluation of the data is as shown in Figure 5.6.

Data collected in an appropriate manner form the basis for statistical evaluation; therefore, the generation of data must be planned carefully. A poorly planned experiment or measurement does not lend itself to statistical analysis. In defining a problem, a number of issues may be considered (90). Table 5.1 outlines these considerations. It is clear from this list that measurements are not conducted in an

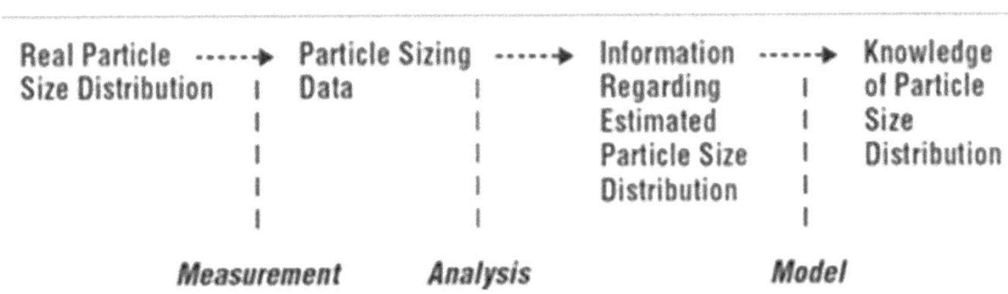

FIGURE 5.6 Sequence of steps involved in evaluating particle size distributions [modified from (90)].

TABLE 5.1
Considerations in Particle Size Technique and Measurements

Initial Consideration	Purpose	Context of Measurement
Desired outcome of measurement (application)	Elucidate nature of particle population	
Parameter to be measured (relevance to application)		
Knowledge of the problem (application)/particle population (facts)		
Assumptions made to initiate measurement (about particles or limitations of technique)	Establish basic nature of the problem (application)	Research and development
		Quality control
		Regulatory compliance
	Temporal nature of problem	Long-term
		Short-term
		Single study
	Spatial nature of problem	Global
		Local
	Prior art	
	Other factors	

information void. Everything that may be known about the system under investigation must be employed to devise, or select, an appropriate measurement technique and procedures for the evaluation of particle size and distribution.

The collection of certain indicators of the quality of data is important. First, the quantitative value of the data must be evaluated. This requires validated methods that have known limits of detection and sensitivity under defined experimental conditions. Quantitative limits for the accuracy and precision of a technique should be stated. The extent to which the data truly represents the population of particle sizes from which it was drawn should be monitored. This often depends upon the selection of appropriate sampling techniques but also depends upon the expert judgment of the individual conducting the measurement. The use of complementary particle-sizing methods will also impact decisions regarding representative sampling.

It is important that the data collected are complete. Certain particle sizing techniques may not measure the entire distribution, which becomes an important factor in the interpretation of the data. A complete analysis of the data will include acknowledgment of the limitations to the methods involved and their impact on the data interpretation.

Further, it is important that the data collected facilitate comparison. This comparison can be made only within a particular method of particle size analysis. Comparisons between differing methods are usually complementary but are not expected to give numerically similar results. All of the issues discussed in the foregoing section become relevant in the use of particle sizing methods as quality control tools (91, 92).

Errors and uncertainty occur at each step described in Figure 5.6. The data collected should be factual—that is, truly representative of a physical reality—or statistical analysis will be meaningless. The errors that can occur in particle size measurement are similar to those that would be expected in any other experimental situation. These errors may be systematic, random, or due to mistakes.

A systematic error is always of the same magnitude and sign and is usually viewed as a bias. Wherever possible, the study should be designed to avoid bias. However, circumstances may occur in which the bias is inherent in the method; such bias should then be acknowledged, and alternative methods considered. Indeed, this was the case with early methods of estimating particle size by using microscopy in which Feret's diameter, for example, was known to overestimate the size of particles. Ultimately, the projected area diameter became the more frequently used descriptor of particle sizes obtained by microscopy. Systematic errors may also occur during data analysis and should be given equal consideration when manipulating data for interpretation.

Random errors occur in any experimental work and are the basis for setting standards on the limits of accuracy and precision. These are handled by the variance considered in statistical analyses.

Mistakes, which may occur in the collection of data or in its analysis and interpretation, do not lend themselves to statistical analysis. A measurement system that is unstable and not in statistical control cannot produce meaningful data (90).

Mathematical Distributions

The description of particle populations requires the adoption of mathematical descriptors of the entire distribution of sizes. Statistical or other mathematical principles have been employed to fit particle size data in an attempt to describe them succinctly.

It is enormously tempting to fit particle size data to a normal or Gaussian distribution of the form:

$$y = \frac{1}{\sigma\sqrt{2\pi}} e^{-(x-\mu)^2/2\sigma^2} \tag{5.13}$$

where μ is the arithmetic mean and σ is the root mean square (RMS) deviation of individual sizes from the average. However, experimental data rarely conform to a normal distribution. Considering the method by which most particles are formed, a normal distribution of particle sizes would not be anticipated. Milling and spraying procedures used to prepare particles are destructive methods that result in the production of small particles from large ones. This is a phenomenon dominated by the volume of particles involved. Since this is a volume function, in broad terms there is a cubic relationship between the large numbers of small particles that can be obtained from small numbers of large particles. However, any method of particle production results in a range of sizes that will not conform strictly to a cubic relationship. Consequently, a number of statistical and mathematical functions have been used to fit particle size distribution data. The range of distributions that may be considered include normal, lognormal, power functions, Rosin Rammler, modified Rosin Rammler, and Nukiyama–Tanasawa (54).

Lognormal distributions are derived from the normal distribution of the logarithms of the particle sizes. Early evaluations of particle size statistics concluded that many populations of particle sizes could be approximated by lognormal mathematical functions. Indeed, the collapse of lognormal distributions into normal distributions was used as a measure of the degree of dispersity of a population of particles. The point of collapse is used to define the practical limit to monodispersity, equivalent to a geometric standard deviation of 1.22 (93).

The advantage of normal and lognormal distributions is that they may be used to extrapolate between different measures of distribution, for example, mass or number. Hatch-Choate equations may be used for this purpose in the case of lognormal distributions. The frequency term for the distribution should be distinguished from individual size parameters. The entire population of particles can be divided into sizes according to specific features such as volume, surface, number, and so forth.

Multimodal Distributions

Complex phenomena may occur based on additive particle size distributions in which multimodal distributions occur. Multimodal distributions require deconvolution if the contributing distributions are to be characterized (94). If a broad, additive distribution occurs, then the overall distribution would appear to fit a normal function according to the central limit theorem (CLT).

Assuming the data have been collected by using a validated method, they may still be easily misinterpreted. Data interpretation depends on the researcher's intent and understanding of the method employed. The first area in which misinterpretation can occur relates to the limitations of the technique used for determining the particle size. Assumptions are often made to conduct an analysis that must be known to allow interpretation of the data. The following list itemizes key examples:

- The ability to examine the entire distribution of particles is a significant requirement. If the distribution extends beyond the working range of the instrument, this should be noted as part of the data interpretation. Ideally, the method will be capable of evaluating the entire distribution.
- The appropriate use of a method is a requirement. If equivalent volume diameters are obtained by laser diffraction, which is an optical method, but the application is to suspensions of particles in a liquid dispersion medium, the data may be of limited relevance to the application. In this example the Stokes diameter would be most relevant to the application.
- A method that meets the first two criteria may be selected, but the sample size may not lend itself to statistical analysis. This is a frequently debated topic with regard to particle sizes derived from microscopic examination, in which the prospect of measuring thousands of particles can be intimidating.

- Finally, the application of a mathematical distribution to fit the practical distribution must be subjected to great scrutiny. The closeness of the fit and, if possible, a physical model to explain the rationale for the fit is desirable.

In the early part of the twentieth century, a major reason for searching for mathematical fits to the data was to allow quantitatively meaningful communication among researchers that would allow studies to be reproduced scientifically or would allow specifications on a product to be set. This should be considered in the historical perspective, in which raw data presented even graphically could be cumbersome. It is no longer as important to fit mathematical distributions to practical data to convey this information easily. Indeed, it is arguably more desirable to give the raw data, without interpretation, to those who wish to reproduce findings or set standards. The importance of mathematical analysis is perhaps most relevant to understanding the principles of particle formation and process reproducibility, in which case the best fit to the data is more important than a single model that fits all particle size distributions. Certainly, the application of the lognormal distribution has been stretched beyond its ability to accurately describe many particle populations. However, presenting raw data is not necessarily as simple a task as it first appears. Some instruments employ algorithms that may not allow optimal visualization of data (95). Indeed, the method that the individual scientist may employ to graphically represent data may be misleading (96). Therefore, caution should be exercised in mathematical, statistical, or graphical representation of the data.

The way forward in presenting data for pharmaceutical purposes lies in the use of complementary techniques. Direct microscopic images of particles help in the interpretation of particle sizing data obtained by other methods. The shape of the particle size distribution and an effective median diameter defined according to the application of the data should be obtained. Where the data application predominantly relates to weight, volume, number, or surface area and is unique for a pharmaceutical process or parameter, the equivalent median diameter (e.g., equivalent weight median diameter for content uniformity, equivalent weighted surface area diameter for dissolution, and equivalent aerodynamic diameter for respiratory deposition) may be weighted in two ways—for the process and for efficiency. These diameters could be suitable for defining specifications and for monitoring the effect of particle size on quality parameters of medicines in development, pharmaceutics, efficacy, or clinical investigations.

It is absolutely essential that particle size data be presented uniquely for each pharmaceutical parameter it is controlling. An inhalation product, for example, has at least three separate reasons for controlling particle size: lung deposition, dissolution upon deposition, and content uniformity and mixing aspects in the initial formulation. In this situation, the overriding factor is lung deposition, but separate calculations need to be performed for the other parameters to understand the effect of working at the bottom and top end of the aerodynamic particle size specification on the content uniformity and dissolution. Even given this analysis, the interpretation does not allow for processing aspects related to the agglomeration of fine particles.

Applications

The assumption of monosized spherical drug and excipient particles in a free-flowing powder for tableting is one of the simplest examples of the application of particle size analysis to the preparation of a dosage form. A random mix can be achieved when the drug and excipient have reached a stable maximum distribution. This is a mathematically random mix because probability is the only factor involved in dispersion. For a completely randomized mix of otherwise identical drugs and excipient spherical particles, the standard deviation, σ_R, has been given by:

$$\sigma_R = \sqrt{\frac{XY}{n}} \tag{5.14}$$

where X and Y are the proportions of drug and excipient particles, respectively, and n is the total number of particles in the tablet. For a completely unmixed system, the standard deviation, σ_0, is given by:

$$\sigma_0 = \sqrt{\frac{X}{Y}} \tag{5.15}$$

A more complex model is required, however, because this is a highly idealized view. Equation (5.16) uses a binomial distribution of particle sizes:

$$\sigma_R = \frac{XY/M}{Y_W\left(1+s^2/w^2\right)_x + X_w\left(1+s^2/w^2\right)_y} \tag{5.16}$$

where X is the fraction by weight of the minor component in the mixture (i.e., the fraction of the drug in a tablet); Y is the fraction by weight of the major component in the mixture (i.e., the fraction of the excipient in a tablet); M is the mass of the sample taken from the mixture (i.e., the weight of a tablet); $w(w_x, w_y)$ is the mean particle weight of the mix components; and $s(s_x, s_y)$ is the standard deviation of the particle weight of the mix components.

Equation (5.14) can be modified, and ultimately equation (5.16) can be rewritten as follows:

$$\sigma_R^2 = \frac{XY/M}{Y\Sigma(fm)_x + X\Sigma(fm)_y} \tag{5.17}$$

where $\Sigma(fm)$ is the mean effective particle weight of the component, which is related to d_{vv}, the volume-weighted, volume mean diameter, as follows:

$$\Sigma(fm) = \frac{1}{6}\pi d_{vv}^3 Q \tag{5.18}$$

The coefficient of variation, C_R, for a random mix, is given by:

$$C_R = \frac{\sigma_R}{X} \tag{5.19}$$

Expression (5.20) is developed by using the Poisson distribution and regarding the variance due to the particle size distribution as a linear combination of several independent randomly distributed variables:

$$C_R = \frac{100\sigma}{X} = 100\sqrt{\frac{\Sigma(fm)_x}{M_x}} \quad (5.20)$$

Use of the Poisson distribution assumes that the random selection of a drug particle as opposed to an excipient particle is a rare event, and it is satisfactory only when the concentration of the drug is low (≤1 percent by weight) and the size of the excipient particles is equivalent to or larger than the size of the drug particles. Substituting the volume-weighted, volume mean diameter, d_{vv}, into the expression for the coefficient of variation (5.20) yields:

$$C_R = 100\sqrt{\frac{\frac{1}{6}\pi d_{vv}^3}{M_x}} \quad (5.21)$$

Taking the logarithm of each side of equation (5.21) shows that for a tablet formulation containing a low concentration of the drug, there will be a linear relationship between log C_R and log d_{vv}, as shown in Figure 5.7.

This relationship is useful in determining the maximum particle size of a drug during formulation design. However, the maximum particle size of a drug can be specified only if a decision is made as to the limiting value of C_R, the coefficient of variation for a random mix. The decision is to some extent arbitrary, and Johnson (1972) recommends a value of C_R of 1 percent or less.

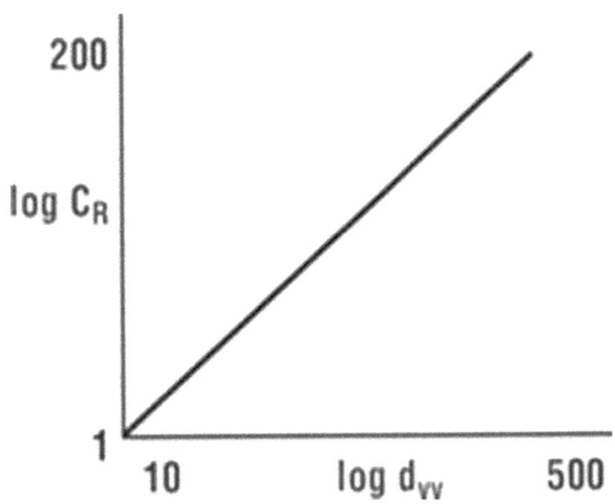

FIGURE 5.7 Theoretical relation between particle size of drug (volume-weighted, volume mean diameter, d_{vv}) and content uniformity of tablets (coefficient of variation for a random mix, C_R).

The characteristic dimension of a particle requires definitions of terms. Since most particles are irregularly shaped, the common practice is to consider equivalent spherical dimensions based on a particular feature such as volume, surface area, or projected area diameters. Populations of particles require numerical data analyses. These data may be represented graphically to convey the particle size distribution. Furthermore, the distribution may be interpreted by statistical or mathematical functions to fit the data. The most frequently employed fit to the particle size distribution is the lognormal function. The parameters derived from this fit are the median diameter and the geometric standard deviation. As a cautionary note, mathematical and statistical functions in their simplest forms may not accurately approximate the data. Indeed, sophisticated approaches are required for multimodal data analysis.

Wherever possible, particle sizing data should be matched to its application. For example, the Stokes diameter, based on sedimentation, is of most relevance to suspensions, and the aerodynamic diameter, based on airborne behavior, is of importance for the deposition of aerosol particles.

Measures of Diameters

Particle diameters can be expressed for a number of characteristics of the particle. These properties include projected area, surface area, volume, and mass. Each of the properties inherent to the particle is an accurate description of the particle, but in any given pharmaceutical process, a particular property will often be the most relevant for predicting the particle's behavior. In some instances, the particle size determined by one method may not be predictive of the particle's performance if the appropriate measurement technique is not chosen. Brittain et al. gave the example of cohesive powders whose flow properties indicated a larger particle diameter than measured by laser diffraction because the particles were dispersed in a liquid before measurement (97).

Diameters determined by optical imaging are one means of describing the primary particle size. Optical methods allow a determination of the particle size and morphology. As an example of the relevance to pharmaceutical processing, the packing and flow properties of a powder are influenced by its morphology. Spherical particles will pack and flow more efficiently than highly irregularly shaped powders. The efficacy of tablet production, then, may best be predicted by diameters that depend upon particle morphology. The ability to increase the density of powders during compaction is the basis for tablet production and is a function of diameter (98). Smaller original particle sizes will result in compacts with increased tensile strength (99). The powder flow properties will be important for the uniform filling of tablet dies (100). Packing properties will affect tablet hardness and, subsequently, dissolution properties (101). Particle morphology is relevant to aerosol dispersion because the flow properties of a powder are determinates of the effective dispersion of powder aerosols from dry powder inhalers (102).

This discussion should make it clear that the characterization of powders for tableting or aerosol dispersion should include some optical measure of the physical diameter and morphology of the powder particles. However, it should also be noted that these measures alone will not be sufficient for optimizing either process. In the case of tableting, dissolution properties will also be a function of the surface area of the powder. For aerosol dispersion, a measure of the aerodynamic diameter is necessary for the prediction of the deposition properties in the lung.

The primary particle size will determine which instrument can be used to provide optical particle size data. Conventional light microscopy may be sufficient for powders intended for tableting, whereas electron microscopy must be used for inhalation aerosol powders because their diameter must be below 5 μm for lung deposition. Although the aerodynamic diameter is the most relevant measure for aerosol dispersion and deposition, this measure alone is not sufficient for the prediction of therapeutic effect. The aerodynamic diameter is related to the geometric diameter through the particle density. Porous particles have recently attracted attention for pulmonary deposition. The particles are engineered to be less than unit density by virtue of a porous structure with trapped air volume within the particle. Because these particles have a mass density significantly lower than unity, particles with an aerodynamic diameter (diameter of a unit density sphere with the same settling velocity as the particle in question) in the respirable size range can be achieved despite having a geometric particle size greater than 10 μm (27). However, the therapeutic effect of the drug particle is determined by the mass of the particle deposited. Therefore, when characterizing aerosol size, a measure of the aerodynamic diameter must also take into account the particle density.

The surface area diameter is another example of a measure that is predictive of powder performance in a desired process. Particle surface area has implications for particle flow, aggregation, and dissolution behavior (47). As stated previously, the disintegration and dissolution properties of a tablet depend on the surface area of the powder in the tablet (101). The dissolving medium may be able to penetrate pores in the particles, if they exist, aiding in disintegration and dissolution. Thus, the measurement of the surface area will be predictive of the dissolution rate. Tablet compact hardness depends on the particle and bonding surface areas of the constitutive powder. These surface areas are functions of the particle size and shape (including the internal surface area of the particles due to their effects on fragmentation) (103). Other studies support the conclusion that the compaction properties of the powder in the tablet depend upon the surface area (104). For particles of very unusual morphology, the interparticulate contact area, measured by nitrogen adsorption, determines the compact hardness of the powder. For the particular powder described in this study, the packing properties could not be ascribed to any other material-related property of the powder.

The volume diameter has relevance to pharmaceutical technology, especially in regard to the delivery of particles in an aerosol. The volume diameter can be measured by electrical resistance techniques, as described earlier in this chapter. As previously described, the therapeutic effect generated by a deposited aerosol particle is determined by the deposited mass, directly related to the volume diameter. Care must be taken, however, for particles with trapped air volume, because the volume diameter will be related to the mass through the particle density. The mass dependence on volume scales with the cube of the diameter, so changes in diameter will have a significant effect on the deposited mass.

In most cases, more than one measure of particle size and characteristics must be collected to predict performance, even if one measure of performance is taken. Foster and Leatherman discussed how multiple measures are needed to describe performance for particles created in a complex process such as spray drying (105). In their study, the morphology and flowability of the spray-dried powders remained the same across multiple equipment parameters during scale-up. Overall performance, however, changed significantly due to changes in particle size, particle size distribution, and bulk density.

PARTICLE SURFACES—SPATIAL AND CHEMICAL DETECTION

A number of techniques are available for probing the composition of the surfaces of materials of pharmaceutical interest. Some of the techniques, such as Auger electron spectroscopy (AES), secondary ion mass spectroscopy (SIMS), and X-ray photoelectron spectroscopy (XPS), can be combined with scanning electron microscopy (SEM) to provide chemical or elemental surface information to complement the direct imaging of the SEM. Another surface technique, gas adsorption, provides surface area information for particles.

Other spectroscopic techniques such as Fourier Transform Infrared and Raman spectroscopy can probe the particle, powder, or solid dosage form for the structure that relates to compatibility with excipients or stability on storage. Moreover, they may be used to probe drug release on dissolution.

AUGER ELECTRON SPECTROSCOPY

The Auger effect occurs due to nonradiative transitions in the inner electron shells of an atom (106). When an energetic particle or photon strikes an atom, an inner shell electron may be ejected. Because this state is unstable, a higher-shell electron will fill the core hole vacancy. The energy difference from this transition will result in either the emission of a photon (usually an X-ray) or the ejection of an electron, called an Auger electron named after Pierre Auger, who studied the effect. The radiative or nonradiative emissions each have a probability of occurring, and the sum of these probabilities must equal one. For exciting electron energies below 10 keV and for atoms with atomic numbers less than 33, the Auger electron emission is favored. The Auger electrons have energies unique to each atom. Therefore, by scanning the energy spectrum, the Auger electron peaks allow the identification of all elements present except hydrogen and helium (107).

Auger electron spectroscopy (AES) has a resolution down to 5 nm. Chemical information can also be obtained by AES but at a reduced spatial resolution. Auger electrons are low-energy particles, so they do not escape from deep in a molecule. Therefore, the sampling depth is limited to about three monolayers. Because an electron beam can be used to create Auger electrons, an Auger microprobe can be constructed by combining SEM with Auger electron detection. This provides a very convenient method to combine spatial and chemical information about particles.

X-Ray Photoelectron Spectroscopy

X-ray photoelectron spectroscopy (XPS) is related to AES, in which the energy of ejected electrons is determined by the measuring instrument. For surfaces irradiated with X-rays, the interaction of a photon with an electron in the atom can lead to the transfer of the photon energy to the electron (108). The photon energy, $h\nu$, must be greater than the electron binding energy, E_B, for the two to interact. Thus, the electron will be ejected with a kinetic energy $E_k = h\nu - E_B$. As for AES, the binding energy is unique to the atomic element, so elemental composition is provided.

Many commercial SEMs combine microscopy with XPS to probe the chemical composition of materials being imaged since SEMs measure emitted electrons. XPS provides information on the chemical bonding of surface materials more efficiently than AES (108). The spatial resolution of XPS is on the order of 5 μm. All elements except hydrogen and helium can be detected. Like AES, XPS is limited to probing the first three monolayers of material. Fine-structure techniques, such as X-ray absorption fine structure (XAFS) and near-edge XAFS can probe the bulk of materials. However, these techniques require high-brightness X-ray sources, such as synchrotrons, and are, therefore, not very practical for pharmaceuticals.

Secondary Ion Mass Spectroscopy

Secondary ion mass spectroscopy (SIMS) involves bombarding a surface with energetic primary ions (0.5–20 keV). Collisions of these ions with atoms in the sample result in the emission of ionized secondary particles that can be analyzed with a mass spectrometer (109). The data can be divided to provide mass and depth information. Unlike XPS and AES, SIMS can obtain bulk information, which provides the ability to detect impurities in the bulk. Spatial resolution is very good, down to 20 nm since an ion beam can be focused by using a magnetic field (107). Marketable instruments are available that scan the ion beam over the sample, much like the SEM, to form an image of mass information.

Gas Adsorption

Gas adsorption studies can be used to determine the surface area of an adsorbing solid (87). Interest in porous particles for aerosol delivery makes this an especially important technique (27). The nature of solid surfaces can be determined by observing the amount of gas adsorbed on a surface as a function of gas pressure. Plots of adsorption versus pressure are called adsorption isotherms. The amount of gas adsorbed in a surface reaction is often proportional to the surface area of the solid (110). At low pressure or at the beginning of adsorption, the isotherm is linear. Clearly, however, the number of adsorption sites on a surface is limited. Brunauer, Emmett, and Teller proposed a set of isotherms based on the simple Langmuir isotherm to allow for adsorption in multilayers (the BET technique) (111). By using different gases, the sizes of surface sites can thus be probed. The BET technique has been used to measure surfaces and describe them in terms of a fractal dimension in which the gas acts as a probe that varies in size (112).

PARTICLE BULK—CHEMICAL PROPERTIES

Fourier Transform Infrared Spectroscopy (FTIR)

The infrared (IR) wavelengths are those in the range 780 nm–1 mm (113). Near infrared wavelengths are in the lower range up to 1 μm (114). IR spectroscopy as an analytical method was developed over 100 years ago. As a qualitative method, the interpretation of IR spectra has been well documented. The development of Fourier Transform IR (FTIR) approximately 50 years ago allowed quantitative interpretation of the spectrum by distinguishing inflections attributable to changes that may have occurred to existing bonds through interactions on storage or with additives.

Traditional IR spectroscopy allows the study of covalent and ionic bonds in organic molecules that can support an understanding of the chemistry of solids and changes that may have occurred during storage or in the presence of additives (115). IR spectroscopy may be employed to identify bond stretches that occur due to interactions between molecules within a solid. These interactions are often correlated with the instability of the product under designated conditions of manufacture or storage. Applying a Fourier Transformation to the plots allows modest differences to be identified and quantified which would be difficult to accomplish by direct comparisons of the data obtained from the original spectrogram of the material (116, 117).

The powder of interest is dispersed in potassium bromide at low concentration and compressed into a disc that can be placed in the path between an IR wavelength source and detector. Samples are prepared of each of the component particles/powders of a formulation and the combination in quantities that represent the final product. Comparisons of individual and combined components allow the interactions that may occur to be observed.

The sensitivity of FTIR has been extended by using attenuated reflectance, rather than absorbance/transmission to collect data (118, 119).

Combining FTIR with differential scanning calorimetry allows measurement of the presence of polymorphs within a single sample which is a valuable indicator of the stability of the formulation (120).

FTIR has a number of applications in pharmaceutical development and quality control. In development, compatibility between components of a dosage form and interactions with the environment (temperature, moisture, and light) can be assessed. In a regulated environment where quality has to be assured, FTIR may be used to identify instability or the presence of impurities.

Raman Spectroscopy

Raman spectroscopy (RS) as originally conceived was developed a century ago. RS is based on inelastic light scattering following photon impingement (121). In current practice, photons from a laser impinging on the sample result in emission at higher or lower energy states, which are indicative of the target's chemistry. Subsequently, the sample is left in a different rotational or vibratory state. RS is a nondestructive chemical analysis technique. It provides detailed information about chemical structure, phase, polymorphism, crystallinity, and molecular interactions (122, 123). It has been used extensively in evaluating the physical chemistry of pharmaceutical and biopharmaceutical products (124, 125).

Raman spectroscopy may be used to explore the morphology beyond the molecular state to the microscale and macroscale related to structured (crystallinity) and unstructured (amorphous) materials identifying the coordination of molecules within the particles.

Thermal Analyses

Thermogravimetric analysis and differential scanning calorimetry are the major forms of thermal analysis (126, 127). Thermogravimetric analysis involves controlled heating of a sample to elevate its temperature while monitoring its mass using a microbalance. The weight loss is then recorded as a function of temperature. This is particularly useful for monitoring moisture loss and release of other volatile materials in addition to identifying the temperature for degradation of the substance being investigated. Differential scanning calorimetry involves a direct comparison of the temperature of a test substance in an aluminum pan to that of a control of simply the pan over a range in which the pans are heated. This technique is useful in identifying phase transition temperatures and the magnitude of energy absorbed to facilitate the transition. As the test substance approaches a phase transition, applied heat is absorbed (endothermic response) by the solid material and used to mobilize atoms or molecules and rearrange them into new crystalline forms (polymorphs) or a new state of matter (liquid). This absorption results in no increase in the temperature of the test substance apparatus with applied heat, whereas the control pan continues to rise in temperature as the heat is applied. Therefore, the difference between the two can be recorded as a developing peak whose area is proportional to the energy absorbed at a transition temperature. It is also possible for heat to be released (exothermic response) with increasing temperature, which would also be detected by difference.

CONCLUSION

Direct imaging of particles provides an irrefutable determination of their size and morphology. Although errors can be made in sizing or morphologies can be altered by sample preparation, direct imaging requires less interpretation of the data than any other method. Some form of direct imaging should be used in combination with any other particle sizing methods to confirm the outcome of those measurements.

The degree of subdivision of a powder impacts on dispersion and sedimentation of suspensions, flow, mixing, and dispersion and dissolution of solid particles. In the case of aerosol products, particle size and distribution also play roles in dispersion, aerodynamic behavior, and deposition.

Particle size measures and their instrumentation are described. It is important to note that measures of particle size cannot be thought of independently from the means of measurement of particle size. Subsequently, these ideas are a step further and tie together particle size and performance in pharmaceutical processes.

Once particles have been prepared, they can be characterized by surface analytical methods to classify their structures and define them prior to their inclusion in pharmaceutical products.

The particle population characteristics area means of predicting performance. These criteria can then be linked to the quality, efficacy, and safety of the product as described in subsequent Chapters 6–10.

Particle morphology is a fundamental element in suspension and powder behavior. An understanding of the principles of particle formation and manufacturing process controls is essential for the optimization of product performance.

This chapter presented a number of particle sizing techniques ranging from direct and indirect imaging of particles to physical methods of separation by size. The choice of method depends on the use of the powder, its physical properties, and its expected size and distribution. As with any process, the cost of measurement, both in terms of monetary and time, is a consideration. The cost of measurement also includes the quantity of powder required by the chosen method for determining the particle size and whether the method is destructive to the sample. This concern will probably be overriding when sizing expensive biotechnology products.

The sizing method employed should be appropriate to the use of powder being measured. For example, the measurement of powders intended for dispersion as aerosols should be measured in their dispersed state. Further, the mechanism used to disperse the powder should be the dispersion method employed for the measurement, if possible. In this example, although geometric diameter may be measured to aid in the quantification of the powder, some means of measuring the aerodynamic diameter of the powder should also be utilized. The aerodynamic diameter is the measure of greatest interest because it will determine the distribution of the powder in the lungs when inhaled as an aerosol. The choice of appropriate statistics is also important. For the pharmaceutical aerosol example, the mass median aerodynamic diameter is the most important measure of size because it is the particle mass that will determine its therapeutic effect.

Also discussed in this chapter is the need to employ more than one particle sizing technique to construct a more complete picture of the product. One of the methods employed should almost invariably be some form of direct imaging. The use of direct imaging serves as a reality check and will ensure that any erroneous data produced by other methods will not go undetected. Many of the more modern techniques of particle sizing, such as laser diffraction, tend to be treated as a black box with unquestioned results. The use of a confirmatory technique is extremely important—at a minimum initially—to validate the instrument's operation.

REFERENCES

1. Allen T. *Particle Size Measurement.* London: Chapman and Hall; 1968.
2. Kaye BH. *Direct Characterization of Fine Particles.* New York: Wiley-Interscience; 1981.
3. ASTM. *Recommended Practice-for Analyses by Microscopical Methods.* Philadelphia, PA: American Society for Testing and Materials; 1973.
4. Watanabe A. Polarizing microscopy of crystalline drugs based on the crystal habit determination for the purpose of rapid estimation of crystal habits, particle sizes and specific surface areas of small crystals. *Yakugaku Zasshi.* 1997;117:771–85.
5. Firestone BA, Dickason MA, Tran T. Solubility characterization of three fluoroquinolone ophthalmic solutions in an in vitro tear model. *Int J Pharm.* 1998;164:119–28.
6. Davidson JA. Arrays of uniform latex spheres as test objects in optical microscopy. *Part Part Syst Char.* 1995;12:173–8.
7. Bastian P, Bartkowski R, Kohler H, Kissel T. Chemo-embolization of experimental liver metastases. Part 1: Distribution of biodegradable microspheres of different sizes in an animal model for the locoregional therapy. *Eur J Pharm Biopharm.* 1998;46:243–54.
8. White JG. In situ determination of delaviridine mesylate particle size in solid oral dosage forms. *Pharm Res.* 1999;16:545–8.
9. Koval M, Preiter K, Adles C, Stahl PD, Steinberg TH. Size of IgG-opsonized particles determines macrophage response during internalization. *Exp Cell Res.* 1998;242:265–73.
10. Cosslett VE, Nixon WC. *X-ray Microscopy.* Cambridge, UK: Cambridge University Press; 1960.
11. Goldstein JI, Newbury DE, Echlin P, Joy DC, Romig AD, Lyman CE, et al. *Scanning Electron Microscopy and X-ray Microanalysis.* New York, NY: Plenum Press; 1992.
12. Danilatos GD. Review and outline of environmental SEM at present. *J Microsc.* 1991;162:391.
13. D'Emanuele A, Gilpin C. Applications of the environmental scanning electron microscope to the analysis of pharmaceutical formulations. *Scanning.* 1996;18:522–7.
14. Hornyak GL, Peschel ST, Sawitowski TH, Schmid G. TEM, STM, and AFM as tools to study clusters and colloids. *Micron.* 1998;29:183–90.
15. Li T, Park K. Fractal analysis of pharmaceutical particles by atomic force microscopy. *Pharm. Res.* 1998;15:1222–32.
16. Stark RW, Drobek T, Weth M, Fricke J, Heckl WM. Determination of elastic properties of single aerogel powder particles with the AFM. *Ultramicroscopy.* 1998;75:161–9.
17. Fuji M, Machida K, Takei T, Watanabe T, Chikazawa M. Effect of wettability on adhesion force between silica particles evaluated by atomic force microscopy measurement as a function of relative humidity. *Langmuir.* 1999;15:4584–9.
18. Wynn EJW, Hounslow MJ. Coincidence correction for electrical-zone (Coulter Counter) particle size analyzers. *Powder Technol.* 1997;93:163–75.
19. Niven RW. Aerodynamic particle size testing using a time-of-flight aerosol beam spectrometer. *Pharm Tech.* 1993;17:72–8.
20. Thornburg J, Cooper SJ, Leith D. Counting efficiency of the API Aerosizer. *J Aerosol Sci.* 1999;30:479–88.
21. Kaye BH, Alliet D, Switzer L, Turbitt-Daouet C. Effect of shape on intermethod correlation of techniques for characterizing the size distribution of powder. Part 2. Correlating the size distribution as measured by diffractometer methods, TSI-Amherst aerosol spectrometer, and Coulter counter. *Particle Particulate Syst Char.* 1999;16:266–73.

22. Hindle M, Byron PR. Size distribution control of raw materials for dry powder inhalers using the Aerosizer and Aero-disperser. *Pharm Tech.* 1995;19:64–78.
23. VanDrunen MA, Scarlett B, Marijnissen JCM. Photon-correlation spectroscopy for analysis of aerosols. *J Aerosol Sci.* 1994;25:347–8.
24. Ma Z, Merkus HG, deSmet JGAE, Heffels C, Scarlett B. New developments in particle characterization by laser diffraction: Size and shape. *Powder Technol.* 2000;111:66–78.
25. Ranucci J. Dynamic plume-particle size analysis using laser diffraction. *Pharm Tech.* 1992;16:109–14.
26. Povey MJW. Particulate characterization by ultrasound. *Pharm Sci Tech Today.* 2000;3:373–80.
27. Edwards DA, Hanes J, Caponetti G, Hrkach J, Ben-Jebria A, Eskew ML, et al. Large porous particles for pulmonary drug delivery. *Science.* 1997;276:1868–71.
28. Yee BM, Sinko PJ. Pharmaceutical Micromeritics. In: Sinko PJ, editor. *Martin's Physical Pharmacy.* Philadelphia, PA: Wolters Kluwer; 2023. p. 528–50.
29. Hickey AJ, Concessio NM. Flow properties of selected pharmaceutical powders from a vibrating spatula. *Particle Particulate Syst Char.* 1994;11:457–62.
30. Hinds WC, Zhu Y. *Aerosol Technology: Properties, Behavior and Measurement of Airborne Particles.* 3rd ed. New York, NY: Wiley; 2022.
31. May KR. A personal note on the history of the cascade impactor. *J Aerosol Sci.* 1982;13:37–47.
32. Reist PC. *Aerosol Science and Technology.* New York, NY: McGraw Hill; 1993.
33. Colbeck I. Dynamic shape factors of fractal clusters of carbonaceous smoke. *J Aerosol Sci.* 1990;21:S43–S6.
34. Fults KA, Miller IF, Hickey AJ. Effect of particle morphology on emitted dose of fatty acid-treated disodium cromoglycate powder aerosols. *Pharm Dev Tech.* 1997;2:67–79.
35. BP. General monographs on aerosols. London, UK: The Stationary Office; 1999.
36. USP. *General Chapter <601> Inhalation and Nasal Products: Aerosols, Sprays and Powders - Performance Quality Tests.* 49. Rockville, MD: United States Pharmacopeia - National Formulary; 2023.
37. US, DHHS, FDA, CDER. *Metered dose inhaler (MDI) and Dry Powder Inhaler (DPI) Products - Quality Considerations. Draft Guidance for Industry.* Silver Spring, MD: US Food and Drug Administration; 2018.
38. Clark AR, Hollingworth AM. The relationship between powder inhaler resistance and peak inspiratory conditions in healthy volunteers - implications for in-vitro testing. *J Aerosol Med.* 1993;6:99–110.
39. Vaughn NP. The Andersen impactor: calibration, wall losses and numerical simulation. *J Aerosol Sci.* 1989;20:67–90.
40. Swanson PD, Muzzio FJ, Annapragada A, Adjei A. Numerical analysis of motion and deposition of particles in cascade impactors. *Int J Pharm.* 1996;142:33–51.
41. Stein SW. Size distribution measurements of metered dose inhalers using Andersen Mark II cascade impactors. *Int J Pharm.* 1999;186:43–52.
42. Mitchell JP, Costa PA, Waters S. An assessment of an Andersen Mark-II cascade impactor. *J Aerosol Sci.* 1988;19:213–21.
43. Esmen NA, Lee TC. Distortion of cascade impactor measured size distribution due to bounce and blow-off. *Am Indust Hyg Assoc J.* 1980;41:410–9.
44. Snell NJC, Ganderton D. Assessing lung deposition in inhaled medications: Consensus statement. Workshop of the British Association for Lung Research, Institute of Biology, London, UK. 17 April 1998. *Resp Med.* 1999;93:123–33.
45. LeBelle MJ, Graham SJ, Ormsby ED, Duhaime RM, Lawrence RC, Pike RK. Metered dose inhalers II: Particle size measurement variation. *Int J Pharm.* 1997;151:209–21.

46. Pezin E, Flament MP, Leterme P, Gayot AT. Method of calibration of twin impinger. *Eur J Pharm Sci.* 1996;4:S142.
47. Hickey AJ. *Pharmaceutical Process Engineering.* 3rd ed. Boca Raton, FL: CRC Press; 2025.
48. Berglund RN, Liu BYH. Generation of monodisperse aerosol standards. *Environ Sci Technol.* 1973;7:147–53.
49. Hickey AJ, Byron PR. Spinning-disk generation and drying of monodisperse solid aerosols with output concentrations sufficient for single-breath inhalation studies. *J Pharm Sci.* 1987;76:60–4.
50. Stockham J, Fochtman E. *Particle Size Analysis.* Ann Arbor, MI: Ann Arbor Sciences; 1979.
51. Stanley-Wood N, Lines R. *Particle Size Analysis.* Cambridge, UK: Royal Society of Chemistry; 1992.
52. Allen T. *Particle Size Measurement.* London, UK: Chapman and Hall; 1990.
53. Piscitelli D, Bigora S, Propst C, Goskonda S, Schwartz P, Lesko L, et al. The impact of formulation and process changes on in vitro dissolution and bioequivalence of piroxicam capsules. *Pharm Dev Tech.* 1998;3:443–52.
54. Dunbar CA, Hickey AJ. Evaluation of probability density functions to approximate particle size distributions of representative pharmaceutical aerosols. *J Aerosol Sci.* 2000;31:813–31.
55. Vaithiyalingam S, Agarwal V, Reddy I, Ashraf M, Khan M. Formulation development and stability evaluation. *Pharm Tech.* 2001;25:38–48.
56. Sacchetti M, VanOort MM. Spray-drying and supercritical fluid particle generation techniques. In: Hickey AJ, editor. *Inhalation Aerosols: Physical and Biological Basis for Therapy.* 2nd ed. New York, NY: Marcel Dekker; 1996. p. 337–84.
57. Tong W-Q, Whitesell G. In situ salt screening-a useful technique for discovery support and preformulation studies. *Pharm Dev Tech.* 1998;3:215–23.
58. Wostheinrich K, Schmidt P. Polymorphic changes of thiamine hydrochloride during granulation and tableting. *Drug Dev Ind Pharm.* 2001;27:481–9.
59. Brittain H. *Polymorphism in Pharmaceutical Solids.* New York: Marcel Dekker; 1999.
60. Brittain H. Using single crystal X-ray diffraction to study polymorphism and solvatochromism. *Pharm Tech.* 2000;24:116–25.
61. Mura P, Faucci M, Parrini P. Effects of grinding with microcrystalline cellulose and cyclodextrins on ketoprofen physico-chemical properties. *Farmaco.* 2001;27:119–28.
62. Suzuki H, Ogawa M, Hironaka K, Ito K, Sunada H. A rifedipine coground mixture with sodium deoxycholate. I: Colloidal particle formation and solid state analysis. *Drug Dev Ind Pharm.* 2001;27:943–9.
63. Parkkali S, Lehto VP, Laine E. Applying image analysis in the observation of recrystallization of amorphous cefadroxil. *Pharm Dev Tech.* 2000;5:433–8.
64. Takeuchi H, Yasuji T, Yamamoto H, Kawashima Y. Temperature- and moisture-induced crystallization of amorphous lactose in composite particles with sodium alginate prepared by spray drying. *Pharm Dev Tech.* 2000;18:355–63.
65. Darcy P, Buckton G. Crystallization of bulk samples of partially amorphous spray-dried lactose. *Pharm Dev Tech.* 1998;3:503–7.
66. Tomas J. Assessment of mechanical properties of cohesive particulate solids. Part 1: Particle contact constitutive model. *Particulate Sci Tech.* 2001;19:95–110.
67. Tomas J. Assessment of mechanical properties of cohesive particulate solids. Part 2: Powder flow criteria. *Particulate Sci Tech.* 2001;19:111–29.
68. Pietsch W. *Size Enlargement by Agglomeration.* New York, NY: Wiley; 1991.

69. Villalobos-Hernandez J, Villafuerte-Robles L. Effect of carrier excipient and processing on stability of indorenate hydrochloride/excipient mixtures. *Pharm Dev Tech.* 2001;6:551–61.
70. Venables H, Wells J. Powder mixing. *Drug Dev Ind Pharm.* 2001;27:599–612.
71. Michoel A, Rombaut P, Verhoye A. Comparative evaluation of co-processed lactose and microcrystalline cellulose with their physical mixtures in the formulation of folic acid tablets. *Pharm Dev Tech.* 2002;7:79–87.
72. Hwang R-C, Gemoules M, Ramlose D. A systematic approach for optimizing the blending process of a direct compression tablet formulation. *Pharm Tech.* 1998;22:158–70.
73. Rietema K. *The Dynamics of Fine Particles.* New York, NY: Elsevier; 1991.
74. Israelachvili J. *Intermolecular and Surface Forces.* New York, NY: Academic Press; 1991.
75. Kano J, Shimosaka A, Hidaka J. A consideration of constitutive relationship between flowing particles. *Kona Powder and Particle.* 1997;5:212–9.
76. Thompson SK. *Sampling.* New York: Wiley; 1992.
77. Orr N. Quality control and pharmaceutics of content uniformity of medicines containing potent drugs with special reference to tablets. In: Deasey P, Timoney R, editors. *Progress in the Quality Control of Medicines.* Amsterdam, The Netherlands: Elsevier; 1981.
78. Kumar V, Kang J, Yang T. Preparation and characterization of spray-dried oxidized cellulose microparticles. *Pharm Dev Tech.* 2001;6:449–58.
79. Palmieri G, Bonacucina G, Martino P, Martelli S. Spray drying as a method for microparticle controlled release systems preparation: Advantages and limits. I Water Soluble drugs. *Drug Dev Ind Pharm.* 2001;27:195–204.
80. Carstensen JT. *Pharmaceutical Principles of Solid Dosage Forms.* Lancaster, PA: Technomic; 1993.
81. Luerkens D. *Theory and Application of Morphological Analysis of Fine Particles and Surfaces.* Boca Raton, FL: CRC Press; 1991.
82. Kaye BH. *A Random Walk Through Fractal Dimensions.* New York, NY: VCH Publishers; 1989.
83. Fayed M, Otten L. *Handbook of Powder Science and Technology.* New York, NY: Van Nostrand Reinhold; 1984.
84. Reimschuessel A, Macur J, Marti J. Microscopy. In: Sibilia J, editor. *A Guide to Materials Characterization and Chemical Analysis.* VCH Publisher; 1988. p. 137–66.
85. Amidon G. Report and recommendation of the USP Advisory Panel on physical test methods - functionality 1: General chapter on particle characterization by microscopy. *Pharm Forum.* 1992;18:4089–92.
86. Green H, Lane W. Sampling and estimation. Particulate clouds, dusts, smokes, and mists. London, UK: Spon Ltd.; 1964. p. 227–83.
87. Grant D. Report and recommendation of the USP Advisory Panel on physical test methods IV. Specific surface area determination of pharmaceutical powders by dynamic gas adsorption. *Pharm Forum.* 1993;19:6171–8.
88. Goldman A, Lewis H. Particle size analysis: Theory and statistical methods. In: Fayed M, Otten L, editors. *Handbook of Powder Science and Technology.* New York, NY: Van Nostrand Reinhold; 1984. p. 1–30.
89. Reist PC. Particle size distributions. In: *Aerosol Science and Technology.* New York, NY: McGraw Hill; 1993. p. 13–30.
90. Taylor JK. *Statistical Techniques for Data Analysis.* Boca Raton, FL: CRC Press; 1990.
91. Shewhart W. *Statistical Method from the Viewpoint of Quality Control.* Mineola, NY: Dover; 1986.
92. Deming W. *Statistical Adjustment of Data.* Mineola, NY: Dover; 1964.

93. Fuchs N, Sutugin A. Generation and use of monodisperse aerosols. In: Davies CN, editor. *Aerosol Science*. London, UK: Academic Press; 1966. p. 1–30.
94. Esmen N, Johnson D. Simulation analysis of inhalable dust sampling errors using multicomponent error model. *Aerosol Sci Tech*. 2001;35:824–8.
95. Lawless P. High-resolution reconstruction method for presenting and manipulating particle histogram data. *Aerosol Sci Tech*. 2001;34:528–34.
96. Menon A, Nerella N. Effective graphical displays. *Pharm Dev Tech*. 2001;6:477–84.
97. Brittain HG, Bogdanowich SJ, Bugay DE, DeVincentis J, Lewen G, Newman AW. Physical characterization of pharmaceutical solids. *Pharm Res*. 1991;8:963–73.
98. Mbali-Pemba C, Chulia D. Lactose texture monitoring during compacting. II: Detailed textural analysis in pore groups. *Drug Dev Ind Pharm*. 1995;21:623–31.
99. Alderborn G. Particle dimensions. In: Alderborn G, Nystrom C, editors. *Pharmaceutical Powder Compaction Technology*. New York, NY: Marcel Dekker; 1996. p. 245–82.
100. Schmidt PC, Rubensdorfer CJW. Evaluation of Ludipress as a "multipurpose excipient" for direct compression. Part I: Powder characteristics and tableting properties. *Drug Dev Ind Pharm*. 1994;20:2899–925.
101. Menon A, Drefko W, Chakrabarti S. Comparative pharmaceutical evaluation of crospovidones. *Eur J Pharm Sci*. 1996;4:S186.
102. Concessio NM. Dynamic properties of pharmaceutical powders and their impact on aerosol dispersion. Dissertation: University of North Carolina at Chapel Hill; 1997.
103. Nystrom C, Karehill PG. The importance of intermolecular bonding forces and the concept of bonding surface area. In: Alderborn G, Nystrom C, editors. *Pharmaceutical Powder Compaction Technology*. New York, NY: Marcel Dekker; 1996. p. 17–53.
104. Rime A-F, Massuelle D, Kubel F, Hagemann H-R, Doelker E. Compressibility and compactability of powdered polymers: Poly(vinyl chloride) powders. *Eur J Pharm Biopharm*. 1997;44:315–22.
105. Foster TP, Leatherman MW. Powder characteristics of proteins spray-dried from different spray-driers. *Drug Dev Ind Pharm*. 1995;21:1705–23.
106. Ferguson IF. *Auger Microprobe Analysis*. New York, NY: Adam Hilger; 1989.
107. Briggs D, Seah MP. *Practical Surface Analysis*. New York, NY: Wiley; 1990.
108. Riviere JC. *Surface Analysis Technique*. Oxford, UK: Clarendon Press; 1990.
109. Wilson RG, Stevie FA, Magee CW. *Secondary Ion Mass Spectrometry: A Practical Handbook for Depth Profiling and Bulk Impurity Analysis*. New York, NY: Wiley-Interscience; 1989.
110. Somorajai GA. *Principles of Surface Chemistry*. Englewood Cliffs, NY: Prentice-Hall; 1972.
111. Brunauer S, Emmett PH, Teller E. Adsorption of gases in multimolecular layers. *J Am Chem Soc*. 1938;60:309–19.
112. Avnir D, Jaroniec M. An isotherm equation for adsorption on fractal surfaces of heterogeneous porous materials. *Langmuir*. 1989;5:143–33.
113. Dufour E. Principles of infrared spectroscopy. In: Sun D-W, editor. *Infra-red Spectroscopy for Food Quality Analysis and Control*. New York, NY: Academic Press; 2009. p. 1–27.
114. Roggo Y, Chalus P, Maurer L, Lema-Martinez C, Edmond A, Jent N. A review of near infrared spectroscopy and chemometrics in pharmaceutical technologies. *J Pharm Biomed Anal*. 2007;44:683–700.
115. Brittain HG. Mid-infrared spectroscopy of pharmaceutical solids. *Profiles of Drug Substances, Excipients and Related Methodology*. 2018;43:321–58.
116. Bunaciu AA, Fleschin S, Aboul-Enein HY. Application of Fourier Transform Infra-red spectrophotometry in pharmaceutical drugs analysis. *Appl Spect Rev*. 2010;45:206–19.

117. Fahelbom KM, Saleh A, Al-Tabakha MMA, Ashames AA. Recent applications of quantitative analytical FTIR spectroscopy in pharmaceutical, biomedical and clinical fields: A brief review. *Rev Anal Chem.* 2022;41:21–33.
118. Salari A, Young RE. Application of attenuated total reflectance FTIR spectroscopy to the analysis of mixtures of pharmaceutical polymorphs. *Int J Pharm.* 1998;163:157–66.
119. van Haaren C, De Brock M, Kazarian SG. Advances in ATR-FTIR spectroscopic imaging for the analysis of tablet dissolution and drug release. *Molecules.* 2023;28:4705.
120. Lin S-Y, Wang S-L. Advances in simultaneous DSC-FTIR microspectroscopy for rapid solid-state chemical stability studies: Some dipeptide drugs as examples. *Adv Drug Del Rev.* 2012;64:461–78.
121. Strachan CJ, Rades R, Gordon KC, Rantanen J. Raman spectroscopy for quantitative analysis of pharmaceutical solids. *J Pharm Pharmacol.* 2007;59:179–82.
122. Taylor LS, Langkilde FW. Evaluation of solid-state forms present in tablets by Raman spectroscopy. *J Pharm Sci.* 2000;89:1342–53.
123. Vankeirsbilck T, Vercauteren A, Baeyens W, Vander Weken G, Verpoort F, Vergote G, et al. Applications of Raman spectroscopy in pharmaceutical analysis. *TrAC Trends in Anal Chem.* 2022;21:869–77.
124. Farquharson S. Pharmaceutical applications of Raman spectroscopy. *Am Pharm Rev.* 2014;17(2).
125. Buckley K, Ryder AG. Application of Raman spectroscopy in biopharmaceutical manufacturing: A short review. *Applied Spectroscopy.* 2017;71:1085–116.
126. Khanna Y. Thermal characterization of materials. In: Sibilia J, editor. *A Guide to Materials Characterization and Chemical Analysis.* 2nd ed. New York, NY: Wiley-VCH; 1996. p. 261–86.
127. Mackenzie R. *Differential Thermal Analysis.* London, UK: Academic Press; 1970.

6 Particle and Powder Behavior

PHYSICAL PROPERTIES

Powder properties may be classified as fundamental (intrinsic) or derived (acquired). Fundamental properties are inherent attributes relating to individual particles or particle populations, such as size and size distribution, shape, porosity, true density, and surface area. Derived properties depend upon the fundamental properties and may be affected by environmental conditions. Examples of derived properties include adhesion and flow.

SIZE

The size of a sphere is uniquely defined by its diameter. However, real particles are rarely homogeneously spherical in nature. The equivalent spherical diameter relates the size of an irregularly shaped particle to the diameter of a sphere having the same size-dependent physical property (1, 2). Typical physical properties measured include those associated with geometry, such as volume, surface, and projected area, and those derived from behavior, such as settling velocity.

Particle sizing techniques may be classified as direct imaging, indirect imaging, physical separation based on particle properties, and charge analysis methods. These methods and instrumentation are described in detail in Chapter 5. The equivalent diameter measured by one method is not directly comparable to sizes measured by other methods due to the fundamental principles involved in each measurement technique. It can be concluded that each method estimates a particular dimension (equivalent diameter) that differs in its relationship to the geometric size in a manner that is not equivalent. Therefore, it would be wise to select the particle sizing method that most closely reflects the product behavior. For example, Stokes' diameter (based on settling velocity) is a logical dimension to measure when considering the stability of particulate suspensions.

A population of pharmaceutical particles rarely contains uniformly sized (monodisperse) particles. Polydisperse particle populations, derived from a single manufacturing process, generally fit a lognormal size distribution and are characterized by the geometric mean size and geometric standard deviation. Statistical analyses of particle populations are discussed further in Chapter 5.

Particle size is a very important parameter for predicting the behavior of powder systems. Particle size greatly affects many derived particle properties, such as adhesion, flow, and fluidization. The fundamental laws of particle motion are strongly

DOI: 10.1201/9781003470151-6

governed by particle size. Monitoring and controlling particle size is a significant element in establishing efficient, reproducible, and cost-effective processes that are the foundation of a quality-by-design strategy in pharmaceutical product development.

Shape

Pharmaceutical particles are rarely spherical. The description of particle shape requires the use of dimensionless numbers that reflect the magnitude of particle irregularity. Shape factors are used to describe deviations from regular geometric shapes, such as spheres and polygons. Other approaches to describing shapes are based on mathematical methods, such as Fourier and fractal analyses.

Shape Factors

Many shape factors have been proposed. The Waddell sphericity factor (ψ) is the ratio of the surface area of a sphere with the same volume as the particle to the surface area of the particle (3). For a sphere, then, $\psi = 1$, and as particles deviate from sphericity, $\psi < 1$. Similarly, the Waddell circularity factor (¢) is the ratio of the circumference of a circle with the same cross-sectional area as the particle to the perimeter of the particle (3). The volume and surface shape factors relate the equivalent diameter to its particle characteristic (3):

$$V = \alpha_v d_v^3 \tag{6.1}$$

and

$$S = \alpha_s d_s^2 \tag{6.2}$$

where V is the volume of the particle, d_v is the equivalent volume diameter, α_v is the volume shape factor, S is the surface area of the particle, d_s is the equivalent surface area diameter, and α_s is the surface shape factor. The surface–volume shape coefficient (α_s/α_v) is the ratio of the surface-to-volume shape factors (1). The value of α_s/α_v equals 6 for spherical particles and exceeds 6 with more asymmetrical particles.

The specific shape factor, α_{sv}, relates particle shape to surface area (3):

$$S = \alpha_{sv}\left(\frac{\omega}{N\rho}\right)^{\frac{2}{3}} \tag{6.3}$$

where S is the surface area, ω is the weight of particles, N is the number of particles, and ρ is the particle density. For spheres, $\alpha_{sv} = 4.85$, and for cubes, $\alpha_{sv} = 6$. For irregular particles, α_{sv} is greater than 6.

The dynamic shape factor, χ, relates the equivalent volume diameter (d_v) to the Stokes diameter (d_s) (4):

$$\chi = \left(\frac{d_v}{d_s}\right)^2 \tag{6.4}$$

The value of χ is 1 for spherical particles and >1 for nonspherical particles.

Other shape factors include the following: Heywood's ratio described particle shape by using the three mutually perpendicular dimensions of length (L), breadth (B), and thickness (T) (1). The elongation ratio is the ratio between the length and breadth. The flakiness or flatness ratio is the ratio between the breadth and thickness. The area ratio (α_a) and prismoidal ratio (p_r) are related to the projected area and volume of the particle, respectively. The values of α_a and p_r enable classification into tetrahedral, prismoidal, subangular, and rounded.

NS is a composite shape factor for two-dimensional particle outlines that considers the deviation from circular, square, and triangular geometric shapes, particle elongation, and the number of corners (5). The NS shape enables classification for a range of different shapes.

Fractal Analysis

Fractal analysis is a mathematical tool used to describe the surface morphology and the degree of surface irregularity of a particle as a single parameter, the fractal dimension (6). The fractal dimension may be determined from either the perimeter or the area of a two-dimensional profile. The concept of fractal geometries is based upon self-similarity at any dimension and relies on repetitive patterns at any scale of scrutiny (7).

For fractal analysis based on particle perimeter, the length of the perimeter depends on the stride length used to measure it. The smaller the stride length, the larger the perimeter measured. A linear plot (known as a Richardson plot) is obtained by graphing the logarithm of the length of the perimeter against the logarithm of the stride length used for the measurements (8). The fractal dimension (δ) is determined by using the following relation:

$$\delta = 1 + |m| \tag{6.5}$$

where m is the slope of the curve of the Richardson plot. The fractal dimension is a description of the space-filling capacity of the profile boundary. Values of δ generally lie between 1 and 2. The higher the value of δ, the more rugged the boundary.

Fractal analysis of the profile of fine particles may produce two linear regions in the Richardson plot. This is due to the different scales of scrutiny to examine the particle. The fractal dimension obtained by using large stride lengths (low resolution) represents the global spatial structure of the particle, whereas the fractal dimension obtained by using small stride lengths (high resolution) represents the detailed textural structure (9).

The use of fractal analysis for characterization of shape in pharmaceutical particles requires the capture of the silhouette boundary profile of the solid particles, generally by scanning electron microscopy with or without image analysis. The image needs to be digitized, and touching particles must be separated with the aid of algorithms before the fractal dimension can be determined. Thus, fractal analysis of individual particles can be tedious and time consuming. Nevertheless, this approach has been used to study the surface geometry of pharmaceutical particles, including excipients (10), diclofenac-hydroxyethylpyrrolidine (11), and β-cyclodextrin-indomethacin (12).

Fourier Analysis

Fourier analysis provides a measure of particle shape using a two-dimensional profile. Particle shape is defined here as the pattern of all the points on the boundary or surface of the particle profile (13). The profile of the particle may be defined as a unique set of x, y coordinates that may be used to regenerate the original two-dimensional shape. Fourier analysis is a mathematical approach used to convert the particle profile into a set of mathematical descriptors. The periodic function for all x values that fits the Fourier equation is the following (14):

$$f(x) = a_0 + \sum_{n=1}^{\infty}\left[a_n \cos(nx) + b_n \sin(nx)\right] = a_0 + \sum_{n=1}^{\infty} A_n \cos(nx - \alpha_n) \tag{6.6}$$

where

$$A_n = \left(a_n^2 + b_n^2\right)^{\frac{1}{2}} \tag{6.7}$$

and

$$\alpha_n = \tan^{-1}\left(\frac{b_n}{a_n}\right) \tag{6.8}$$

The variable n is the harmonic order, and the Fourier coefficients A_n and α_n are the harmonic amplitude and harmonic phase angle, respectively.

Three different methods of Fourier analysis are used to characterize particle profiles (13). The radial method converts the particle profile to polar coordinates of the radial length from the center to a boundary as a function of angle (R, θ). However, this method is not applicable to concave-shaped particles (14). The Granlund descriptors are obtained from the boundary coordinates in the complex plane. This method is widely used to describe particle shapes. The increased elongation and squareness of disodium cromoglycate particles following the treatment with fatty acids have been assigned numerical values by using this method (15). The Zahn-Roskies (ZR) descriptors are obtained from the angular bend as a function of arc length. The ZR method classifies particles into five shape categories (perfect square, square-shaped, elongated, rounded, and very irregular) (14).

Surface Roughness

Surface roughness of particles may be characterized using various types of profilometry, including stylus tip profilometry, laser profilometry, and atomic force microscopy (AFM) (16).

In stylus tip profilometry, a fine, sharply pointed stylus tip, diamond or sapphire, is drawn across the sample surface. The vertical movement of the tip as it traces the surface irregularities provides the height of the surface. The resolution of measurement is dictated by the size of the tip and the aperture angle. The standard tip size is around 5 μm, which enables measurement of similar-sized valleys, or castellations. Surface roughness may be examined as contour lines across the sample, or as three-dimensional areas consisting of several lines. However, during the tip motion, soft surfaces may be damaged and larger asperities may be removed. AFM provides higher-resolution three-dimensional images, in the nanometer range, by using smaller tip sizes. Using AFM, the surface topography is scanned with the sharp tip, similar to stylus tip profilometry. No surface treatment or coating is required, and samples may be imaged in ambient air or within a liquid environment (17). The use of a "tapping mode," in which the tip is moved up and down between individual sites, enables the examination of soft samples. Due to the vertical range limitations of AFM, difficulty arises in the measurement of height variations >10 μm. Laser profilometry provides a noncontact method by which the sample surface is tracked by using a laser beam with a spot size of around 0.6–1 μm. The relative depth of the beam spot at the sample surface is determined by using two-beam interferometry.

Surface roughness parameters include rugosity, root mean square deviation, and maximum peak-to-valley height. The rugosity (R_a), also known as the center line average (CLA) height, is the arithmetic mean average distance of all points of the profile from the center line, given by (16):

$$R_a = \frac{1}{n}\sum_{i=1}^{n}|y_i| \tag{6.9}$$

where n is the number of points and y_i is the distance of the point i from the center line.

The root mean square (RMS or R_q) deviation describes the variability of the profile from the center line (16):

$$R_q = \sqrt{\frac{1}{n}\sum_{i=1}^{n}y_i^2} \tag{6.10}$$

The maximum peak-to-valley height is the difference between the maximum and minimum points of the profile.

Fractal analysis may be used to examine the particle profile determined by profilometry, enabling the surface irregularity to be characterized by using the fractal dimension.

Surface Area

The specific surface area is the surface area per gram of the sample, and it may be determined by either the gas adsorption method or the permeation method, mercury intrusion porosimetry. The gas adsorption method determines surface area by the amount of gas or liquid that forms a monolayer on the sample surface. The permeation method determines surface area by the rate at which a gas or liquid permeates a powder bed.

The degree of gas adsorption by a solid depends upon the chemical nature of the solid and gas molecules, the temperature, and the partial pressure of the gas. Adsorption isotherms plot the amount of gas adsorbed on a surface as a function of gas pressure. Various types of adsorption isotherms exist (2). Type I isotherms, described by the Langmuir equation, are linear when the logarithm of the amount of gas adsorbed is plotted with the logarithm of gas pressure. Type II isotherms are sigmoidal-shaped isotherms, in which a monolayer is formed by the first inflection, and multilayer formation occurs with continued adsorption. Type IV isotherms are produced by gas adsorption onto porous solids. Isotherms of Types III and V are not common and are produced when the heat of adsorption in the monolayer is less than the heat of condensation on subsequent multilayers. The size of surface adsorption sites can be probed by different-sized gas molecules and described in terms of a fractal dimension (18).

The most common adsorbing gas for surface area determination is nitrogen, and the adsorption isotherm is generally Type II, as described by the Brunauer, Emmett, and Teller (BET) equation (19):

$$\frac{P}{Q(P_0 - P)} = \frac{1}{Q_m C} + \frac{(C-1)}{Q_m C} \cdot \frac{P}{P_0} \tag{6.11}$$

where Q is the volume of nitrogen adsorbed, P is the pressure, P_0 is the saturation vapor pressure of liquid nitrogen, Q_m is the volume of gas required to form a monolayer at normal temperature and pressure, and C is a constant related to the difference between the heat of vaporization of a monolayer and the heat of vaporization of liquid nitrogen.

A linear fit is obtained by plotting $P/[Q(P_0 - P)]$ against P/P_0, which allows Q_m and C to be determined graphically. The specific surface area is calculated from these values. This technique assumes that a close-packed monolayer of adsorbed nitrogen is formed, and the cross-sectional area of adsorbed molecules is the same as the closest packing in a solidified gas.

Density

Various definitions and measurements of density are used. True density is the density of the actual solid material, exclusive of voids and intraparticle pores larger than the molecular or atomic dimensions in the crystal lattice. True density is determined by using liquid displacement or helium pycnometry. Granule density is exclusive of pores <10 μm and is determined by using mercury displacement. The bulk density of a powder is the mass of a powder occupying a unit volume. Bulk density includes voids and intraparticle pores (2). It is a function of true density and packing arrangement and is also referred to as apparent density.

The packing of particles may be described in terms of void space (20). The porosity (ε) is the fraction of total void volume and is determined from the relationship:

$$\varepsilon = 1 - \frac{\rho_b}{\rho} \qquad (6.12)$$

where ρ_b is the bulk density and ρ is the true density. In the "rhombohedral" arrangement, which is the closest packing for uniform-sized spheres, each sphere contacts 12 other spheres and has a porosity of 25.95 percent (21). In the "cubical" arrangement, which is the most open packing, each sphere contacts six other spheres. This arrangement has a porosity of 47.64 percent. Intermediate arrangements include "orthorhombic" and "tetragonal spheroidal." Real particles have porosity values between 30 and 50 percent.

Bulk density is affected by the particle size distribution, shape, and adhesion properties. Tapping or vibration causes rearrangement of the particles, reducing void space and increasing bulk density. In powders with wide-size distributions, smaller particles may fill the void spaces between the larger particles. As particles deviate from sphericity, porosity increases. Increased void space may be observed with rough particle surfaces. Smaller particles may produce increased porosity due to cohesion. When particles are packed into a cylinder, the voids along the wall are greater than in the body of the bed. In cases where the particle diameter is very small compared with the wall diameter, these wall effects become appreciable.

The bulk density of a powder becomes important in pharmaceutical processes that subdivide the powder by volume. For example, variations in bulk density result in weight and dose variation of tablets.

PARTICLE ADHESION

Two opposing forces are involved in particle interactions: attraction and detachment. Adhesion is the result of the attractive interaction between two unlike particles, and cohesion is the result of the attractive interaction of similar particles. Detachment forces consist mainly of gravitational and acceleration forces. Both detachment forces depend on mass; thus, the size of the adherent particles will influence the degree of particle interaction.

The many types of adhesion forces may be classified as electrical and nonelectrical. The total force of adhesion at any time is equal to the sum of the individual forces (22).

Electrical Forces

Contact potential forces and Coulombic forces are the electrical interactions existing in the air; double-layer interactions occur in aqueous liquids.

Contact Potential Forces

Contact potential forces occur between adjacent uncharged and unlike particles by electron transfer due to differences in the energy state of the Fermi levels, the outermost conduction band of electrons. Electron transfer occurs upon contact between particles until equilibrium is achieved, which results in a contact potential difference. Surface charges may be measured following detachment of the particle. The magnitude of the contact potential force (F_e) may be calculated by (23):

$$F_e = \frac{2\pi q^2}{S} \tag{6.13}$$

where q is the particle charge on detachment from a substrate, and S is the contact area between the particle and the substrate. Contact potential forces depend on the sizes and contact areas of the particles and the electron donor-acceptor properties of the materials. The acquisition of electrostatic charges on particles from contact with other particles or surfaces is generally referred to as triboelectric charging. It may occur during such processes as sieving, mixing, milling, and pneumatic conveying.

Coulombic Forces

Coulombic forces occur between charged particles and an uncharged surface. The charged particle induces an equal and opposite charge on the surface, producing image forces. The Coulombic interaction is situated on both sides of the surface at equal distances from the surface. The adhesion force results from the image interaction between particle charges and the induced charge in the substrate. The adhesion force resulting from Coulombic interactions (F_{im}) may be determined by (23):

$$F_{im} = \frac{Q^2}{l^2} \tag{6.14}$$

where Q is the particle charge and l is the distance between the centers of the charges. As a result of charge leakage, the magnitude of Coulombic interactions is reduced over time. The rate of charge decay may be exponential or bi-exponential and depends on relative humidity, temperature, and the presence of adsorbed layers.

Electrostatic Double Layer

Particles in liquid media develop surface charges by selective adsorption of ionic species in the liquid; by ionization of surface groups, such as carboxylate and amine groups; or as a result of differences in the dielectric potential constant between the particle and the liquid (24, 25). The surface charge is balanced by counterions of opposite signs, thereby forming a double layer that surrounds the particle. The closest layer to the particle surface consists of tightly bound solvent molecules and is called the Stern layer. The outer layer (diffuse layer) contains partially mobile ions due to greater solvent mobility. The boundary between the two layers is called the shear plane.

The Derjaguin–Landau–Verwey–Overbeek (DLVO) theory describes the interaction between charged particles in liquids (26). As similarly charged particles approach each other, the electrostatic repulsion must be overcome before the separation distance is sufficiently close to allow particle attraction due to van der Waals forces. Increased adhesion is observed with oppositely charged particles.

The zeta potential is the potential difference between the shear plane and the electroneutral region of bulk liquid (24, 25). It is affected by the ionic concentration, pH, viscosity, and dielectric constant of the solution and has practical application in the stability of liquid dispersion systems.

Electrostatic Charge Measurement

Electrostatic charge measurement in air is performed using the Faraday well/cage or electrical low-pressure impactor. Electrostatic charge measurement in liquids is performed by electrophoresis.

Faraday Well/Cage: The electrostatic charge of particles is measured by an electrometer after pouring bulk powder into a Faraday well or following particle removal from a Faraday cage in a moving airstream (27, 28). The electrostatic charge of particles is expressed as a specific charge for each gram of powder. The electrostatic charge polarity and magnitude were determined from various pharmaceutical excipients following triboelectrification in a cyclone and from detached drug particles in model interactive systems (27, 28). Modification of the Faraday well enabled the determination of the electrostatic charge of aerosolized particles (29).

Electrical Low-Pressure Impactor: Electrical low-pressure impactors (ELPIs) allow the aerodynamic size classification and electric charge measurement of aerosol particles in the 0.03–10 μm range (30). The electrical charge of particles is determined as they impact onto collection stages using a multichannel electrometer.

Electrophoresis: The zeta potential of particles in liquid media is determined by electrophoresis, which measures the rate of particle movement in an electric field. In an applied electric field, charged particles move toward the oppositely charged electrode. Particles with greater charge move faster. The particle velocity may be characterized by electrophoretic mobility, which is the ratio of the velocity of the particles to the field strength. The zeta potential may be calculated from the particle mobility by using the Smoluchowski equation:

$$\zeta = \frac{\eta\mu}{\varepsilon} \tag{6.15}$$

where ζ is the zeta potential, η is the viscosity of the medium, μ is the particle electrophoretic mobility, and ε is the dielectric constant.

The speed of particle movement during zeta-potential measurements was originally determined manually with a microscope but is now automated. Newer methods use laser Doppler velocimetry (LDV), in which the Doppler frequency shift provides a measure of particle speed.

The stability of colloidal dispersions depends on the zeta potential of particles. Formulation of stable dispersions should aim at absolute values of zeta-potential above 50 mV, which provide mutual electrostatic repulsion of particles (31). Coagulation and sedimentation occur when the zeta potential is close to zero.

Nonelectrical Forces

Nonelectrical interactions comprise intermolecular forces, capillary forces, and forces due to solid bridge formation and mechanical interlocking.

Intermolecular Forces

Intermolecular forces include interactions predominantly due to van der Waals forces, such as dipole–dipole, dipole-induced dipole, and induced dipole–induced dipole interactions. Hydrogen bonding is another type of intermolecular force involving the interaction between a highly electronegative element, such as oxygen in a hydroxyl group of one molecule, and hydrogen from another hydroxyl group.

Van der Waals forces are present in gaseous and liquid environments. Particles may be brought into contact by electrostatic forces, but once adhesion has been established, van der Waals forces keep particles in contact, and other forces are required for their detachment (32). Electrostatic and capillary forces are smaller in magnitude than van der Waals forces. The van der Waals forces (F_{vdW}) are described by (32, 33):

$$F_{vdW} = \frac{Ar}{6z^2} \text{ for a sphere and a plane surface} \tag{6.16}$$

$$F_{vdW} = \frac{Ar}{12z^2} \text{ for two identical spheres} \tag{6.17}$$

where A is the Hamaker constant, r is the particle radius, and z is the distance between the sphere and surface. The Hamaker constant depends upon the molecular properties of the materials involved. Van der Waals forces decrease rapidly with separation distance and are effective only at separation distances smaller than 10 Å (34). The magnitude of van der Waals forces becomes negligible compared with that of gravitational forces when the particle size exceeds 10 μm (Figure 6.1) (33).

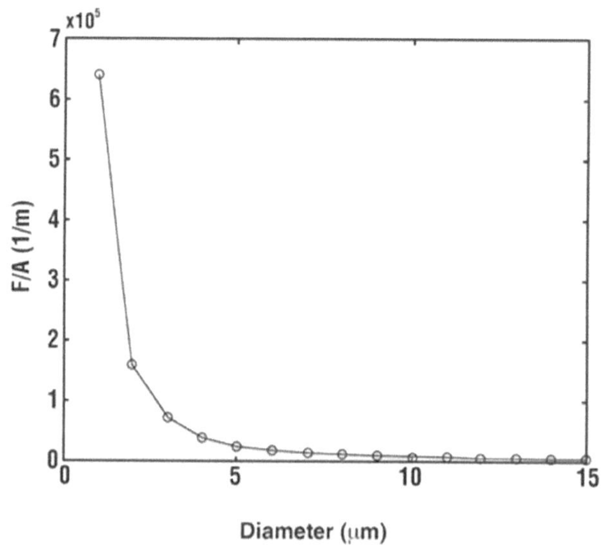

FIGURE 6.1 Influence of particle diameter on the magnitude of van der Waals forces.

Capillary Forces

Capillary forces occur following condensation of water vapor between particles, which causes liquid bridge formation (35). The resultant surface tension produces attractive forces. The capillary force (F_c) between a spherical particle and a planar surface may be described as (23, 36):

$$F_c = 4\pi r\gamma \cos\theta \quad (6.18)$$

where r is the particle radius, γ is the surface tension, and θ is the contact angle between the particle and the liquid.

Liquid bridge formation occurs over time; thus, capillary forces are not exhibited immediately after contact of particles to substrates. Factors affecting the magnitude of capillary interactions include relative humidity (RH), particle size, and surface roughness. Generally, in conditions below 50 percent RH, capillary forces do not produce significant adhesion; however, capillary forces will be the dominant adhesion force between 65 percent and 100 percent RH (23).

Solid Bridges

Solid bridges between particles may form by chemical reaction, partial melting followed by subsequent solidification, or partial dissolution followed by subsequent crystallization (22, 35). Solid bridge formation occurs over time and reinforces other types of preceding interactions. The formation of solid bridges following drying depends on hygroscopic properties (37).

Adhesion Force Measurements

The force of adhesion may be determined by the force required for particle detachment because this force is equal in magnitude but opposite in sign to the adhesion force (38). Methods developed for measuring adhesion forces involve either individual or bulk particle detachment. Techniques that use single particles involve the direct measurement of the force required for detachment of each particle. Techniques that use multiple particles involve the determination of a number of particles detached following the application of a specific force.

Single-Particle Detachment Methods

Historically, single-particle detachment methods determined the forces of adhesion for relatively large particles (38). The adhesion forces between fine particles were too small to detect (35). However, advances in sensitivity have enabled the measurement of forces between surfaces at much higher resolution (39). Adhesion force may be measured between two spheres or a sphere and a plane surface. Methods for determining adhesion force using single-particle detachment include microbalance, pendulum, and atomic force microscopy.

Microbalance Method: The microbalance method measures the adhesion force of a spherical particle attached to the end of a quartz fiber, which acts as a cantilever, helical spring, or torsion balance (40). The force of adhesion is measured by the elongation of the quartz spring as follows (38):

$$F = c\Delta h \tag{6.19}$$

where c is a parameter characterizing spring rigidity, and Δh is the elongation of the spring at the moment of detachment (Figure 6.2). The preliminary compression force producing particle adhesion affects the force of interaction due to differences in true contact area between particles (38). The microbalance method has been used to determine the adhesion force of simulated quartz and Pyrex™ particles to various surfaces, where increased particle size or humidity increases adhesion, and increased surface roughness reduces adhesion (41).

Pendulum Method: The pendulum method determines the adhesion force from the detachment angle of a freely hanging sphere acting under gravitational force. Detachment occurs in a perpendicular direction to the contact area, and the adhesion force is determined from the angle of detachment as follows (38):

$$F = mg\sin\alpha \tag{6.20}$$

where α is the angle of deviation from the vertical, m is the mass, and g is the gravitational acceleration (Figure 6.2).

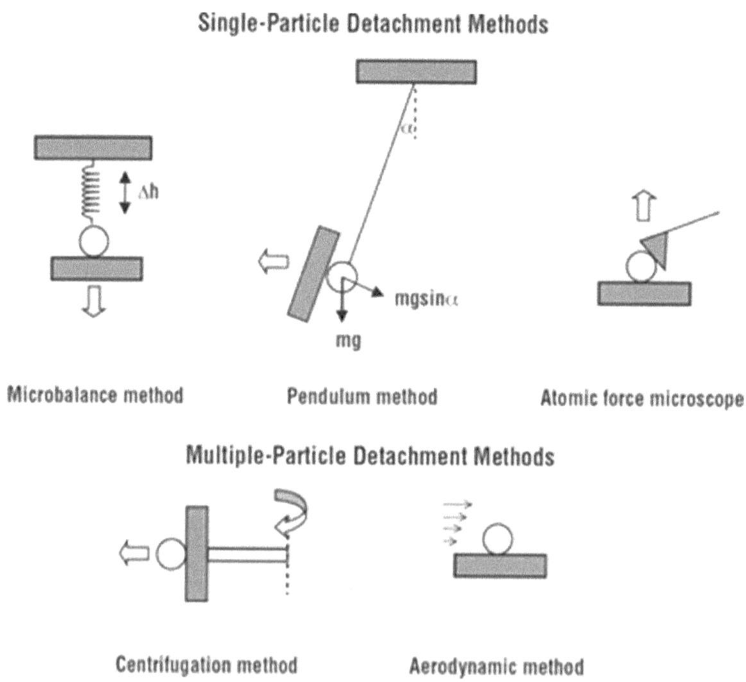

FIGURE 6.2 Methods for measuring particle adhesion.

Atomic Force Microscopy: The atomic force microscope (AFM) (42), originally developed as a high-resolution imaging tool, may be used for adhesion force measurements. Adhesion forces acting between the AFM probe and sample produce vertical displacement of the probe cantilever. As the sample surface is brought into and out of contact with the probe, the cantilever deflection is plotted as a function of the separation distance between the probe and sample. This plot is known as a force-distance curve (43). The adhesion force is calculated from these force-distance plots, according to Hooke's law (17, 43):

$$F = kx \tag{6.21}$$

where k is the cantilever spring constant, and x is the vertical displacement of the cantilever. Equation (6.21) is similar to equation (6.19) for the microbalance. The vertical deflection is measured in the AFM technique, whereas the spring elongation is measured in the microbalance method. Accurate calibration of the cantilever is required to provide quantitative force measurements.

The use of the colloidal probe technique allows measurement of the adhesion force between a spherical colloid particle attached to the tip of the AFM cantilever and the sample surface (Figure 6.2) (44). The colloid probe technique has been used to measure adhesion forces between particles and surfaces in air and in liquids (17, 44). The use of a spherical colloidal particle provides a well-defined geometry for sample contact (45). Particle adhesion is dependent on the surface roughness of either the particle or the surface and the contact geometry between the probe and sample. Increased adhesion may occur when a particle fits snugly in pits or grooves of the surface, whereas decreased adhesion may occur when particles sit atop ridges or bumps on the surface (46). A higher applied load may result in higher adhesion due to increased contact area.

AFM studies have been used to investigate the adhesion force between spheres or particles. For example, Fuji et al. observed increased adhesion between two silicon spheres in humid conditions (70 percent RH) due to capillary condensation (47). Sindel and Zimmerman measured the adhesion forces between two lactose particles, and between a lactose particle and compressed lactose and sapphire substrates, in both a vacuum and air (48). Louey et al. determined the adhesional properties of lactose carriers for dry powder inhalers using a colloid probe (49). The adhesion forces at individual sites were stable, and the lognormal force distributions obtained enabled characterization by the geometric mean force and geometric standard deviation. The sensitivity of the technique was sufficient to differentiate between the samples examined. The ability to characterize surfaces of drug delivery systems using AFM has the potential to advance the understanding of the relationship between surface properties and delivery system functionality (17).

Multiple-Particle Methods

Methods for measurement of bulk-particle detachment forces include the use of centrifugation, vibration, aerodynamic force, and impaction. All of these techniques involve the detaching force acting simultaneously on all adherent particles present on the surface.

Centrifugation: The centrifuge method determines adhesion following particle detachment from the adhering surface during rotation around an axis (Figure 6.2). The magnitude of the detaching force (F) is determined by the following relation (38):

$$F = V(\rho - \rho_0)(j + g) \tag{6.22}$$

with

$$j = \omega^2 l \tag{6.23}$$

and

$$\omega = \frac{2\pi n}{60} \tag{6.24}$$

where V is the particle volume, ρ and ρ_0 are particle and surrounding media densities, respectively, j is the centrifugal acceleration, g is the gravitational acceleration, ω is the angular rotating velocity of the surface, l is the distance between the surface and the axis of rotation, and n is the number of revolutions of the surface per minute. For adhesion in air, where $\rho >> \rho_0$ and $j >> g$, the force of detachment may be simplified to (38):

$$F = V\rho\omega^2 l \tag{6.25}$$

The adhesion force distribution is obtained by measuring the number or mass of detached particles at various centrifuge speeds. It generally follows a lognormal relationship, enabling characterization by the geometric mean force (F_{50}) and the geometric standard deviation (GSD) or by the mean centrifuge speed (S_{50}; the speed required to detach 50 percent of adherent particles) and GSD (38, 50, 51). Particle detachment may be determined optically (52–54) or by chemical analysis (51, 55, 56). The centrifuge speed must be increased or decreased smoothly to eliminate the influence of inertial forces, and each speed should be held for several seconds. However, increasing the centrifuging time does not have any effect on particle detachment in the air.

The centrifuge technique determines adhesion forces between pharmaceutical particles and substrate surfaces or compressed powder surfaces (54, 57–59). Increased centrifugal forces obtained by the ultracentrifuge enable the determination of the adhesion force between micronized particles and compressed powder surfaces (60). Detached drug particles were separated from intact interactive units by using a wire mesh or a brass plate holder (50, 51, 55, 56, 61). However, distortion of the mesh occurred at high rotation speeds; in addition, disintegration and flattening of the carrier were encountered (62). The detachment angle was not always 90°, as expected for the ideal sphere resting on a planar surface. Particles located in clefts on the carrier surface were proposed to roll or slide along a surface and dislodge other particles in their path before detachment (62). Particle-on-particle adhesion was examined by attachment of interactive mixtures to an aluminum support disk (63).

A modified centrifuge technique determined the frictional forces between particles and surfaces by mounting the surface at angles between 0° and 90° (64). The force of detachment (F_{det}) decreased with decreasing angle of detachment (α). A linear relationship was obtained by plotting F_{det} cos α against F_{det} sin α. The coefficient of friction (μ) and the frictional force (F_{frict}) were determined from the slope and intercept, respectively.

Vibration: The magnitude of the detaching force (F) by vibration can be determined by the following relation (38):

$$F = m(j + g) \tag{6.26}$$

with

$$j = 4\pi^2 v^2 y \cos\left(wt + \frac{\pi}{2}\right) \tag{6.27}$$

and

$$w = 2\pi v \tag{6.28}$$

where m is the particle mass, j is the vibrational acceleration, g is the gravitational acceleration, ν is the vibration frequency, y is the vibration amplitude, and t is time. The detachment force is adjusted by varying the vibration frequency and amplitude. More particles are detached by the vibration method at relatively small forces of detachment than by the centrifuge method. However, centrifugation detaches more strongly adhered particles than does vibration. Particle deformation may occur with the vibrational force; thus substrates subject to plastic deformation are not suitable for this technique (38).

The vibration technique has been employed to determine the adhesion force and segregation tendency of powder mixtures (65–71). The vibration apparatus consists of either stacking cylinders or a single sieve. The dislodged drug particles pass through the sieve aperture, whereas the intact interactive unit is retained above the sieve. However, the comminution of carrier particles enables undetached drug particles to pass through the sieve (68).

Aerodynamic Force: An adherent particle may be dislodged by an airstream flowing across the surface (Figure 6.2) when the aerodynamic force provided by the drag (F_{dr}) meets or exceeds the force of adhesion (F_{ad}), particle weight (P), and lift of the airstream on the particle (F_{lif}) (72):

$$F_{dr} \geq \mu(F_{ad} + P + F_{lif}) \tag{6.29}$$

where μ is the coefficient of friction.

The aerodynamic force of drag may be described by the relation (72):

$$F_{dr} = \frac{C_x \rho v^2 \pi r^2}{2} \tag{6.30}$$

where c_x is the drag coefficient, ρ is the air density, v is the airflow velocity, and r is the particle radius. The adhesion force distribution is obtained by measuring particle detachment at various airflow velocities. The distribution follows a lognormal function, enabling characterization by the median detachment velocity and geometric standard deviation (72). This technique is also referred to as the hydrodynamic method (73). The detachment velocity is dependent on the particle size and shape and the surface roughness.

Impaction: The impact technique determines the adhesion force from the detachment of particles by the impact of a hammer or bullet from the opposite face of the adhering surface. The magnitude of the detaching force (F) can be determined by the simple force relation (74, 75):

$$F = ma \tag{6.31}$$

where m is the impacting mass and a is the impact acceleration.

A pendulum impact technique using a swinging hammer was employed to determine the adhesion force between organic and pharmaceutical particles and a substrate surface (74, 75). The impaction force was adjusted by changing the lift angle of the impact hammer and the hammer weights. The number of particles detached after impaction was determined. A lognormal force distribution was obtained, enabling characterization by F_{50}, the separation force at which 50 percent of the particles remained on the substrate (74). The effect of particle shape, surface roughness, and the presence of fine particles on the adhesion force was examined by this technique (76). The median particle size of excipients that remained adhered to the substrate surface was correlated to powder flow characteristics (75, 77).

Factors Affecting Particle Adhesion

Particle adhesion is affected by fundamental particle properties, such as particle size and shape and surface roughness, and environmental factors, such as storage duration and humidity.

Increasing the particle size causes an increase in adhesion force (41). However, the forces of detachment (acceleration and gravitational forces) are dependent on particle mass. Thus, smaller particles have relatively greater forces of adhesion as compared with forces of detachment due to their reduced particle mass (40). Gravitational forces become negligible at particle diameters $<10\ \mu m$, and adhesion forces become more dominant (32). Particle detachment is proportional to particle diameter (d) for the microbalance method, whereas detachment force is proportional to d^3 for the centrifugation and vibration methods and to d^2 for the aerodynamic method (40). Greater particle detachment with increased particle size has been observed when using centrifugation (50, 78, 79).

Theoretical equations for adhesion assume that particles are ideally smooth and spherical. Particle shape and surface roughness may affect particle adhesion due to their effect on contact area and separation distance. Surface roughness may increase the separation distance between particles, thus reducing van der Waals forces, which

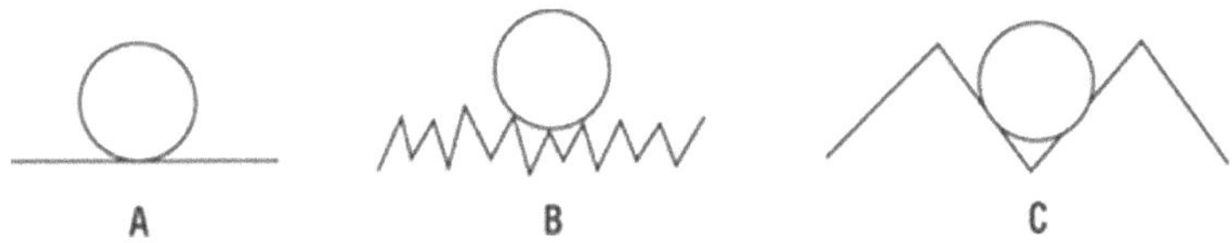

FIGURE 6.3 Types of surface roughness affecting particle adhesion: (A) ideally smooth, (B) microroughness, and (C) macroroughness. Adapted from (80).

become negligible at approximately 10 Å (34). Surface roughness may be described as ideally smooth, or having microroughness and macroroughness (Figure 6.3) (80). Compared to ideally smooth surfaces, reduced contact area and adhesion occur when particles contact the top of the asperities on surfaces with microroughness. Conversely, increased contact area and adhesion occur when particles contact the valley between two asperities on surfaces with macroroughness. The effect of surface roughness depends on the size of the asperity, the interval distance between asperity peaks, and the relative size of the adhering particle (81). Increased physical stability of interactive mixtures containing macrorough carrier particles was found to be due to multiple adhesion sites and entrapment of the adherent particles in surface clefts, which prevents detachment caused by rolling, abrasion, or collision with other particles during further mixing or processing (82). Indeed, surface roughness may increase interparticulate forces by mechanical interlocking (83). Increased adhesion of salicylic acid particles on Emdex™, Dipac™, and lactose carriers was attributed to greater surface roughness (56).

The orientation of nonspherical particles adhered to a surface may alter the contact area. Plate-shaped particles may have increased adhesion due to their flat contact area. Increased adhesion was observed as particles became more spherical, an effect attributed to increased contact area (76). In contrast, the reduced adhesion observed in particles with surface protuberances or adhered fine particles was attributed to reduced contact area (76).

The chemical composition of particles affects their deformity, strength, and electrical properties (65). The chemical and electrical nature of particles affects their adhesion due to the influence of these factors on van der Waals, electrostatic, and capillary forces (40). In addition, the viscoelastic properties of the particles influence the degree of surface deformation and contact area of adhesion (40). Plastic deformation causes permanent deformation and increases adhesion. Conversely, elastic deformation produces reversible deformation and may indirectly affect adhesion due to its influence during particle detachment. Contact areas increase due to deformation as time of contact is extended. Particle adhesion with elastic deformation may be described by using the Derjaguin–Muller–Toporov (DMT) or Johnson–Kendall–Roberts (JKR) models. The DMT model is applicable for hard materials with Young's modulus greater than 10^9 N/m^2 and considers the interaction as a ring-shaped zone around the contact area. The JKR model is more appropriate for soft materials and assumes the interaction force exists at the contact area only (32, 84).

Differences in polymorphic state and degree of crystallinity may affect adhesion due to differences in physicochemical properties. Capillary forces depend upon the moisture adsorption of the contacting materials and the time required for liquid bridge formation. Increased adhesion after storage in humid conditions was observed in mixtures that contained carriers with high moisture affinity, due to capillary interactions, whereas no change was observed in carriers with minimum moisture sorption (37.). The crystallization of amorphous regions may result in solid bridging and permanent agglomeration of particles (85).

Electrostatic charge influences particle adhesion directly through electrostatic interactions and indirectly by increasing van der Waals interactions due to reduced separation distance. Oppositely charged drug and carrier particles facilitate close surface contact and increase the strength of van der Waals forces (82). The stability of binary and ternary interactive mixtures was affected by the electrostatic charges of the adhering particles (55).

Environmental factors, such as mixing and storage duration and conditions, may affect the adhesional properties of powders. Increased adhesion occurred in interactive mixtures with increased blending times (86). Possible explanations for this phenomenon include increased triboelectric charging, the presence of drug aggregates at shorter blending times enhancing drug detachment, greater intermolecular interaction due to increased impact and surface deformation with time, or fracture of particles with increased blending time enabling adhesion of new surfaces.

Applied force during particle contact may affect particle adhesion. Increasing the applied force increased particle adhesion due to the larger contact area caused by plastic deformation (54, 57). Plastic deformation was material-dependent and was related to yield pressure (87). Increased adhesion occurred with increased duration of applied pressure, due to the time dependence of plastic deformation (88). An increase in temperature increases the plasticity of materials, thus enhancing adhesion (89). Larger particles produced increased contact area and adhesion with plastic deformation under applied force (90). Greater increases in adhesion occurred with increasing applied force for irregular-shaped particles compared to spherical particles (54).

During storage, the electrostatic charge decreases due to a process known as "charge decay," thus reducing the total adhesional force (23, 27, 91). The rate and extent of charge decay were dependent upon the relative humidity and particle system. For example, the rate of charge decay increases with increasing relative humidity (92). In addition, at increasing relative humidity, van der Waals forces increase because the adsorbed moisture layer produces reduced distances between particles.

Increased adhesion was observed following storage above 65% RH because capillary condensation formed a liquid layer (liquid bridge) in the gap between the contacting bodies (93). Increased adhesion due to capillary forces also occurred following storage at high humidity (5, 37, 94). Solid bridge formation was observed following oven drying (100°C) or column-drying of mixtures stored at high relative humidities (37, 95).

PARTICLE MOTION IN BULK POWDERS

Particle motion in bulk powders may be observed in processes such as powder flow, fluidization, and mixing as described in the following sections.

Powder Flow

Powder flow requires the expansion of the packed bed, individual particles to overcome interparticle forces, and increased particle separation distance, all of which combine to allow particle mobility. The flow of powders may be characterized by four phases—plastic solid, inertial, fluidization, and suspension (96). At rest, powders exhibit a plastic solid phase, characterized by small interparticle spacing and slow particle velocities. The powder bed stresses are independent of velocity. When the powder is tilted above its angle of repose, a transition to the inertial phase occurs. In the inertial phase, the interparticle spacing increases, particle collisions produce powder bed stresses, and powder flow begins but is confined to the surface layer. The inertial phase is the powder flow characterized in pharmaceutical powders. When the interstitial fluid provides sufficient momentum to overcome interparticle forces, fluidization occurs. During fluidization, the interparticle spacing is of the order of the particle size. Very large interparticle spacing, negligible interparticle forces, and particle velocities similar to the entraining fluid velocity characterize the suspension phase.

Factors affecting powder flow relate to those affecting particle adhesion and cohesion (83). These include particle size, shape, porosity, density, and surface texture (2). Powder flow is reduced as particle size decreases, due to the greater adhesion forces relative to gravitational forces. Generally, particles greater than 250 μm are free-flowing, particles smaller than 100 μm have reduced flow, and particles smaller than 10 μm are very cohesive and have poor flow under gravity, except as large agglomerates (83). The presence of particles smaller than 10 μm promotes cohesiveness and may reduce powder flow (97). The flow properties of such powders may be improved by removing the "fines" or adsorbing them onto added larger particles (2). Particle shape influences powder flow due to the contact area between particles. Optimal flow properties are provided by spherical particles, which have minimal interparticulate contact. Irregular-shaped particles, such as flakes or dendrites, have high surface-to-volume ratios and poorer flow properties (83). Particle density affects powder flow through its influence on the relative contributions of gravity and surface forces. Denser particles generally have better flow properties (97). Surface roughness affects powder flow due to its influence on particle adhesion. Rough particles have a greater tendency for mechanical interlocking than particles with smooth surfaces (83). Moisture adsorption on the particle surface may improve powder flow by the combined effects of pore-filling, increased particle density, and lubrication to a certain level, whereas capillary forces produce reduced flow (97).

Methods for measuring powder flow are based on static bed or dynamic flow properties, as discussed next.

Static Methods

Static bed methods include angle of repose, angle of spatula, Carr's compressibility index, Hausner ratio, and shear cell measurements.

Angle of Repose and Angle of Spatula

When a bulk powder is allowed to fall freely and accumulate on a flat surface, it forms a cone of material. The angle of repose—the angle that the surface of the cone makes with the horizontal surface—has been stated to be an indirect measure of particle size, shape, porosity, cohesion, fluidity, surface area, and bulk (98). It provides an indication of the ease at which the powder will flow. Small angles of repose (less than 40°) indicate free-flowing powder, whereas large angles (greater than 50°) indicate poor flow properties. However, the angle of repose does not provide any indication of the flow pattern or flow rate. The angle of repose increases as the particle departs from sphericity and as the bulk density increases. It is independent of particle size when the particle diameter is larger than 100 μm, but it increases sharply with particle diameters smaller than 100 μm. The value of the angle of repose can differ, depending on which of the various measurement methods are used.

The angle of spatula is similar to the angle of repose. It is the angle of powder remaining on a spatula that has been vertically removed from a powder bed. An average angle is measured prior to and after gentle tapping. For a given material, the angle of spatula is generally higher than the angle of repose (98). Similar to the angle of repose, smaller values are indicative of better flow.

Carr's Compressibility Index and the Hausner Ratio

Powder flow properties may be empirically estimated by using the porosity and packing density of powders. Carr's compressibility index and the Hausner ratio are examples of these estimates.

Carr's compressibility index empirically estimates the powder flow from packing density and compressibility. Carr's compressibility index (CI) is based on the difference between the tapped and bulk density, and is expressed as a percentage:

$$\mathrm{CI} = \frac{\rho_{\mathrm{tap}} - \rho_{\mathrm{bulk}}}{\rho_{\mathrm{tap}}} \cdot 100\% \tag{6.32}$$

where ρ_{tap} is the tapped density and ρ_{bulk} is the bulk density of a powder. Compressible powders are expected to be more cohesive and exhibit poor flow; thus, powders with lower CI values are estimated to have better flow behavior. Powders with CI values >20 percent have been found to exhibit poor flow properties (98).

The Hausner ratio estimates powder flow based on packing density and compressibility, similar to Carr's compressibility index. The Hausner ratio (HR) is calculated by:

$$\mathrm{HR} = \frac{\rho_{\mathrm{tap}}}{\rho_{\mathrm{bulk}}} \tag{6.33}$$

Jenike Shear Cell

A shear cell apparatus was developed to characterize powder flow by measuring the shear stress at different values of normal stress (99). The powder is placed under a normal load until plastic consolidation occurs. A horizontal shear force is applied at reduced consolidation loads, and the force required for shear movement is measured (100, 101). Various parameters are obtained from the yield locus, which is defined as the relationship between the shear strength of a material and the applied load. The equation for the yield locus is (100):

$$\frac{\tau}{\tau_p} = \left(\frac{\sigma + S}{\sigma_p + S}\right)^{\frac{1}{n}} \tag{6.34}$$

where τ_p is the shear strength of the powder bed under the consolidation load σ_p; τ is the shear strength of the powder bed under a reduced load σ; S is the tensile strength of the powder; and n is the shear index. The tensile strength (S) is the negative load at which the shear strength is zero, and the shear index (n) is a measure of the degree of curvature of the yield loci. Values of n lie between 1 and 2; n approaches 1 for free-flowing powders and approaches 2 for poor-flowing powders. Other parameters obtained from the yield locus include the angle of internal friction (φ), which is a measure of the shear strength of material per unit of consolidation force, and the unconfined yield stress (f_c), which is a measure of the stress required to cause a failure in shear when the material is unsupported in two directions. The angle of internal friction is equal to the angle of repose.

Dynamic Methods

Dynamic flow methods include the rotating powder drum and the vibrating spatula.

Rotating Drum

As a powder sample is slowly rotated in a drum, the powder rises until its angle of repose is exceeded and an avalanche occurs. The avalanching behavior is characterized by either the time interval between avalanches or the angle of repose. Although an identical rank-order correlation of flow properties was determined in a study using time-to-avalanche data, better differentiation between powders was obtained by using the angle of repose data (102).

The dynamic angle of repose oscillates around a mean value. The change of angle of repose with respect to the mean angle plotted against time produces an oscillating data plot. The change of angle of repose with respect to the mean angle plotted against the mean angle of repose produces a phase space attractor plot (103), where data points cluster around a central attractor point. The scatter around this attractor point represents a measure of variability in the avalanching behavior. Free-flowing powders have a tight attractor, whereas cohesive powders have widely fluctuating attractor plots. Lower attractor points indicate better flowability (104). Oscillating data and phase-space attractor plots may also be used as indicators of deterministic chaos, by which periodic, random, and chaotic patterns can distinguish different powder behaviors. Mathematical analysis is used to derive fractal dimensions and Hausdorff, or capacity dimensions (103).

Vibrating Spatula

The vibrating spatula characterizes dynamic powder flow by use of the fractal dimension (105, 106). The mass of powder flowing from a vibrating spatula is measured as a function of time. Fractal analysis is performed on the mass versus time profile, and the fractal dimension is determined. Generally, fractal dimension values range between 1 and 2, with lower values indicative of more regular flow behavior. This method has been used for the characterization of lactose, sodium chloride, and maltodextrin particles (105, 106).

Fluidization

As a fluid (gas or liquid) flows upward through a particle bed at a low flow rate, the fluid percolates through the interstices of the fixed bed. As the flow rate increases, the bed expands as the particles move apart until they become suspended in the fluid. The incipient fluidization point occurs when the drag force exerted by the fluid counterbalances the gravitational force on the particles. In liquid systems, smooth progressive bed expansion occurs as the flow rate increases, and particles are homogeneously distributed throughout the bed. In gas systems, fluidization depends on the particle properties and interparticulate forces. The capacity of powders to be fluidized by gas depends on the rate at which they expand (107). Higher gas fluidization rates produce greater expansion and better powder flow, unless particle cohesion hinders the penetrating gas, producing slow dilution and bubbling.

Gas fluidization of powders has been empirically characterized into four groups by using the Geldart classification (Figure 6.4) (108).

Group A powders are slightly cohesive powders. At low gas velocities, they are homogeneously fluidized in a stable expanded state (107). At higher gas velocities, heterogeneous fluidization occurs with rapid mixing and bubbling of gas through the powder bed, similar to boiling liquid. The bubble size is affected by particle size, proportion of fine particles (smaller than 45 μm), pressure, and temperature. The powder bed undergoes aeration and large powder bed expansion (108).

Group B powders exhibit negligible interparticle forces. Fluidization occurs with bubbling, slightly above the incipient fluidization. The bubbles burst at the surface as discrete entities. Minimal aeration and powder bed expansion occur. The powder bed collapses rapidly when the fluidization gas is removed. Little or no powder circulation occurs in the absence of bubbles (108).

Group C powders are cohesive and extremely difficult to fluidize, due to the stronger interparticle forces compared with the fluidizing force. The particles are generally smaller than 20 μm, soft or irregular-shaped (108). Channelling and plugging occurs at velocities greater than the incipient fluidization gas velocity, and the gas escapes through vertical channels in the powder bed while the remainder of the

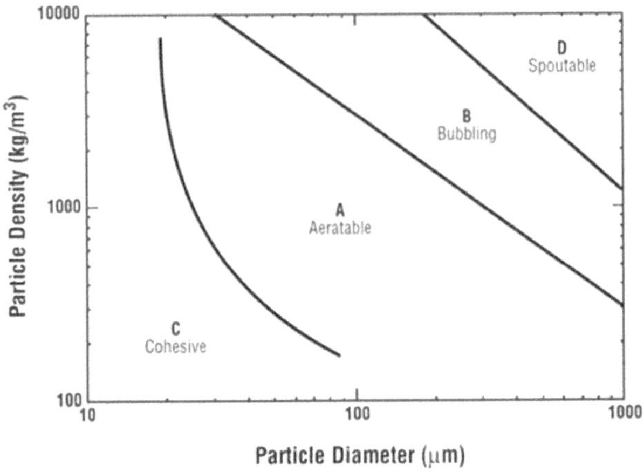

FIGURE 6.4 Geldart fluidization classification of powders. Adapted from (108).

powder bed remains stagnant with very little expansion. This action is attributed to the high cohesion between individual particles (107).

Group D powders consist of large or dense particles. During fluidization, spouting of the fluidizing gas occurs. The bubbles rise more slowly than the interstitial fluidizing gas, causing the gas to flow into the base of the bubble and out of the top. This results in relatively poor mixing of solids. Size segregation of particles is likely to occur (108).

Mixing

Powder mixing depends on the particle interactions. Random mixing occurs between noninteracting particles, whereas interactive mixing involves particle adhesion.

Random Mixing

Random mixing occurs with statistical randomization, in which the probability of finding one type of particle at any point in the mixture is equal to the proportion of that type of particle in the mixture. Mixing and demixing processes occur simultaneously. The rate of mixing follows first-order decay. The three principal mechanisms involved in mixing include diffusion, shear, and convection (109). Diffusion produces the redistribution of particles by random movement of the particles relative to one another, analogous to gaseous diffusion. Shear mixing involves the three-dimensional movement of adjacent particles due to the formation of slip planes within the powder bed. Convection produces the movement of groups of adjacent particles to another part of the powder bed. Diffusion is referred to as micromixing, whereas shear and convection are referred to as macromixing. Factors affecting random mixing include particle size, shape, density, friability, and mixing proportions. Segregation in random mixtures may occur by percolation and densification, by which smaller particles percolate through void spaces, due to agitation and vibration or gravity, and produce caking at the bottom of the powder bed under the force of the weight above. Segregation may also occur with particle acceleration. Different-sized particles have different stopping distances when projected in a horizontal direction, which may result in trajectory segregation.

Interactive Mixing

Interactive mixing generally involves the adhesion of small particles (usually the drug) onto the surface of a larger carrier particle (usually the excipient). Interactive mixtures, originally known as ordered mixtures, produce stable homogeneous mixes due to particle adhesion (110). During mixing, high energy is required to break up the cohesive forces within agglomerates of small particles (111). Factors affecting particle adhesion within interactive mixtures include the size, shape, and surface morphology of both drug and carrier particles, drug concentration, and the presence of ternary components. Two types of segregation mechanisms exist for interactive mixtures. Constituent segregation occurs when the adhered drug particles become detached from carrier particles. Interactive unit segregation occurs when interactive units of different sizes undergo percolation or trajectory segregation.

Mixing equipment may be classified as batch or continuous mixers (109). The entire powder sample is blended and removed in batch mixers. The energy input varies greatly between mixers. Low-energy units, such as V-shaped blenders, use rotation of the shell for mixing. High-energy units, such as ribbon blenders, use rotating blades to break up powder aggregates. Energy input is required to overcome the cohesive forces within particle aggregates so the separate particles can move in relation to one another (112). Low-energy mixers may not be able to provide sufficient energy for the blending of cohesive and electrostatic powders. In these cases, the agglomerates rather than the individual particles are mixed.

PARTICLE MOTION IN GASEOUS DISPERSIONS

The dynamics of particles in gaseous systems are described in different size regimes: the free molecule regime for particles smaller than 0.01 μm, the transition regime for particles in the 0.01–0.4 μm range, the slip flow regime for particles larger than 0.4 μm, and the continuum or Stokes' regime for particles larger than 1.3 μm (4).

Free Molecule Regime

Particles in the free molecule regime undergo thermal diffusion due to Brownian motion as a result of the small particle size compared to the mean free path of gas molecules. Brownian motion is the constant random movement of small particles suspended in a fluid caused by their constant bombardment by fluid molecules. Each particle moves with uniform velocity, with smaller particles moving more rapidly than larger ones. Particles in high concentrations move more rapidly than those in dilute concentrations. Also, particles move more rapidly through media with lower viscosity. The amplitude of motion is directly proportional to the absolute temperature. The diffusion of the particles depends on the motion of the fluid (113). These particles remain suspended due to the gas convective motion and the high particle diffusion force, which is more significant than the gravitational settling force (114).

Continuum (Stokes') Regime

Particle movement in the continuum size regime follows Stokes' law. A particle falling freely under gravitational forces accelerates until a terminal velocity is attained. The terminal settling velocity (v) in laminar conditions is given by Stokes' law:

$$V = \frac{d^2 g(\rho_s - \rho_0)}{18\eta} \tag{6.35}$$

where d is the particle diameter, g is the gravitational acceleration, ρ_s is the particle density, ρ_0 is the density of the dispersion medium, and η is the viscosity of the dispersion medium. Stokes' law assumes that particles are rigid spheres, the fluid is incompressible, the fluid velocity at the particle surface is zero, and the particle motion is constant and not affected by other particles or nearby walls.

The nature of motion (laminar, intermediate, or turbulent) depends on the particle size, velocity, and properties of the surrounding fluid. The Reynold's number, Re, is a nondimensional number that describes the nature of motion (115):

$$Re = \frac{\rho_0 dv}{\mu} \tag{6.36}$$

where ρ_0 is the density of the fluid; d is the particle diameter; v is the relative velocity between the particle and fluid; and μ is the coefficient of viscosity. Re represents the ratio of inertial forces to frictional forces acting on the particle. Laminar flow is observed in spherical particles with Re values smaller than 1. As Re values increase above 1, eddies form on the downstream side of the particle and the flow is turbulent. Particle shape and orientation during settling may profoundly affect the drag force acting on the particle and settling velocity. The nature of flow in pipes is also determined by Re. The boundary of Re values between laminar and turbulent flow differs between flow around particles and within pipes.

The resistance to motion is due to the opposing drag force of the particle. The drag force increases with particle size and deviations from sphericity. The coefficient of drag, C_D, is a correctional factor used to account for deviation in Stokes' law due to particle size and shape factor. The value of C_D is dependent on Re (116):

$$C_D = \frac{24}{Re} \quad \text{for} \quad Re < 0.1 \tag{6.37}$$

$$C_D = \frac{24}{Re} + 4.5 \quad \text{for} \quad 0.1 < Re < 10 \tag{6.38}$$

and

$$C_D = \frac{24}{Re} + \frac{4}{Re^{\frac{1}{3}}} \quad \text{for} \quad Re > 10 \tag{6.39}$$

For particles with Re values greater than 1000, $C_D \cong 0.44$, and particle movement follows Newton's law, of which Stokes' law is a special case (117). The relationship between particle shape and the coefficient of drag has been established by Carmichael (118).

Slip Flow Regime

Small particles produce a faster-settling velocity than expected from Stokes' law. This is attributed to the slip of fluid molecules at the particle surface, resulting in less resistance encountered. Various correction factors have been applied to Stokes' law to correct the slip flow effect and to extend the lower particle range. The Cunningham slip factor (C_c) extends the Stokes' law region to 0.1 μm (117):

$$C_C = 1 + \frac{2.52\lambda}{d_p} \tag{6.40}$$

where λ is the mean free path of gas molecules and d_p is the particle diameter.

The slip correction factor (C) extends the range of Stokes' law to below 0.01 μm (119):

$$C = 1 + \frac{\lambda}{d_p}\left[2.514 + 0.800\exp\left(-0.55\frac{d_p}{\lambda}\right)\right] \tag{6.41}$$

Values of C are close to unity for particles larger than 1 μm. Larger values are observed as the particle size decreases. Tabulated values of C for spheres of unit density are available (119).

Centrifugal Settling

The rate of particle settling in a centrifugal field is based on Stokes' law, where the gravitational acceleration is replaced by the centrifugal acceleration (120):

$$V_r = \frac{d^2\omega^2 R(\rho - \rho_0)}{18\eta} \tag{6.42}$$

where v_r is the centrifugal settling velocity, R is the radius of the circular path, ω is the angular velocity of the particle, ρ is the density of the particle, ρ_0 is the density of the fluid d is the particle diameter, and η is the fluid viscosity.

Centrifugation is generally used for sedimentation of extremely fine particles. Particles may be suspended in air or in a liquid medium. Cyclone separators utilize radial acceleration to separate particles suspended in a gas stream (120). The dust-laden gas enters the cyclone tangentially, causing a vortex that descends into the cone. The circular motion imparted on the airstream causes the suspended matter to move toward the outer walls and fall into a collecting bin located below. A greater separating force is required to move smaller particles from the airstream to the outer wall. The tangential velocity of the particle is assumed to be close to that of the airstream.

Acceleration, Deceleration, and Curvilinear Motion

When a particle undergoes straight-line acceleration under the influence of constant or varying forces, its velocity depends upon the particle's mass and mobility. The relaxation time, τ, is the time required for the particle's velocity to adjust to the new condition of forces (121):

$$\tau = mB \tag{6.43}$$

where m is the particle mass and B is the mechanical mobility of the particle. Constant velocity is considered to be reached when $t = 3\tau$, where t is the time since acceleration/deceleration was applied. Particle motion usually occurs over much longer time periods. However, it may be assumed that a particle reaches terminal velocity instantaneously.

A particle will continue to travel after an external force is removed, depending on the initial velocity and the relaxation time of the particle. The stopping distance (S) in the absence of external forces is given by (121):

$$S = \tau v_0 = mBv_0 \tag{6.44}$$

where v_0 is the initial velocity.

In a moving gas stream passing around an obstacle, the particles follow a curved path. Smaller particles with negligible inertia follow the gas streamlines. Large and heavy particles may impact the obstacle due to inertia. Curvilinear motion is characterized by Stokes' number, which is the ratio of the stopping distance of a particle to the size of the obstacle (121). As Stokes' number approaches zero, the particles follow the gas streamlines perfectly, and as Stokes' number increases, the particles resist changing their original direction. The collection of particles by inertial impaction is used widely for size fractionation of aerosols, as discussed further in Chapter 5.

PARTICLE MOTION IN LIQUID MEDIA

The movement of particles within a liquid dispersion occurs by diffusion or gravitational sedimentation, depending on particle size. Field-flow fractionation, a method used to separate particles from liquid media, will be discussed in the following section. The rheology of liquid dispersions will also be discussed in the following section.

Diffusion

Particles diffuse spontaneously by Brownian motion from regions of higher concentration to regions of lower concentration until equilibrium is reached. Brownian motion is a random movement caused by the bombardment of particles by the liquid molecules. Higher particle velocities and faster diffusion arise with smaller particle sizes. Brownian motion is reduced by increased viscosity of the dispersion medium. The rate of particle movement due to diffusion is described by Fick's first law of diffusion (24, 25):

$$\frac{dM}{dt} = -DS\frac{dC}{dx} \tag{6.45}$$

where dM/dt is the rate of mass movement, dC/dx is the concentration gradient, D is the diffusion coefficient, and S is the area of diffusion.

Gravitational Sedimentation

Stokes' law may be used to describe the rate of gravitational settling for dilute suspensions containing 0.5–2 percent solids concentration. However, Stokes' law does not hold true for suspensions with higher solids concentrations, which exhibit hindered settling because the particles interfere with one another (122).

Brownian motion is more significant with colloidal particles, and gravitational settling is negligible. A stronger force, such as centrifugal acceleration, is required for the sedimentation of colloidal dispersions. In coarse suspensions of particles smaller than 2–5 μm, Brownian motion counteracts gravitational sedimentation, producing stable suspensions. Although the particle size is greater than 0.5 μm, these coarse suspensions may exhibit Brownian motion at low viscosity.

Field-Flow Fractionation

Field-flow fractionation (FFF) is an elution technique capable of simultaneous separation and measurement of macromolecular, colloidal, and particulate materials, ranging from about 1 nm to more than 100 μm. The separation occurs by differential retention in a stream of liquid flowing through a thin channel (50–300 μm) with an applied gradient field at right angles to the flow (123). The applied field drives particles into different stream laminae, and the unequal velocities of the laminae result in the separation of particles along the flow axis. Particles moving in the flow laminae closest to the wall are more slowly entrained. The separated components are eluted one at a time into a detector. The observed retention time is related to various physicochemical properties, such as mass, size, density, and electrical charge.

Sedimentation FFF employs the spinning of the channel to generate differential acceleration forces at right angles to the flow. These forces are highly selective in separating colloidal and larger particles. The single-particle force (F) in sedimentation FFF is:

$$F = V_p \Delta\rho G \tag{6.46}$$

where V_p is the particle volume, $\Delta\rho$ is the difference in density between the particle and carrier liquid. and G is the field strength or acceleration.

Flow FFF employs two right-angle flow streams. The channel stream sweeps components toward the outlet. The crossflow stream, entering and exiting through permeable walls, drives components toward the accumulation wall. The driving force is the viscous force exerted on a particle by a crossflow stream. Other examples of applied fields used in FFF include temperature and electrical fields.

Flow of Liquid Dispersions

Rheology describes the flow of liquid dispersions. Liquid systems are classified as either Newtonian or non-Newtonian. The rate of shear is proportional to the shearing stress in Newtonian systems (Figure 6.5) according to the following relation (124):

$$\frac{dv}{dr} = \eta \frac{F}{A} \tag{6.47}$$

where dv/dr is the rate of shear; F/A is the shearing stress; and η is the coefficient of viscosity, commonly known as viscosity, and defined as the resistance of a system to flow under applied stress. Higher temperature produces increased viscosity of gases and decreased viscosity of liquids. As the viscosity increases, greater applied forces are required to produce flow. The viscosity of colloidal systems depends on the particle shape of the dispersed phase. Spherical particles produce lower viscosity than linear particles.

The three types of non-Newtonian flow behavior are plastic, pseudoplastic, and dilatant (Figure 6.5). In plastic flow, flow begins only when a certain shear stress, known as the yield value, is exceeded. At shear stress less than the yield value, elastic deformation occurs. At shear stress greater than the yield value, Newtonian behavior is exhibited. Plastic flow is associated with flocculated particles in concentrated suspensions. Higher yield values indicate a higher degree of flocculation. Systems exhibiting plastic flow may be characterized by using the yield value and the plastic viscosity. Materials exhibiting plastic flow are known as Bingham bodies.

In pseudoplastic flow, increasing shear stress produces an increased rate of shear and a reduced viscosity. Pseudoplastic flow is exhibited by polymers, due to linear alignment of molecules with increasing shear stress. This alignment reduces the internal resistance and increases the shear rate. Pseudoplastic systems are also known as "shear-thinning" systems.

In dilatant flow, increasing shear stress produces an increase in viscosity. Dilatant systems increase in volume under applied shear force. Dilatant systems are also known as "shear-thickening" systems. When the shear stress is removed, the system returns to its original state of fluidity. Dilatant systems generally contain a high solids concentration (>50 percent) of small-deflocculated particles. At rest, particles are closely packed with minimum void space filled with fluid. At increasing shear stress, the particles need to take an open packing form to move past each other, resulting in expansion or dilation of the system. The volume of fluid then becomes insufficient to fill the increased void space. Increased resistance to flow results because the particles are no longer completely wetted or lubricated by the fluid.

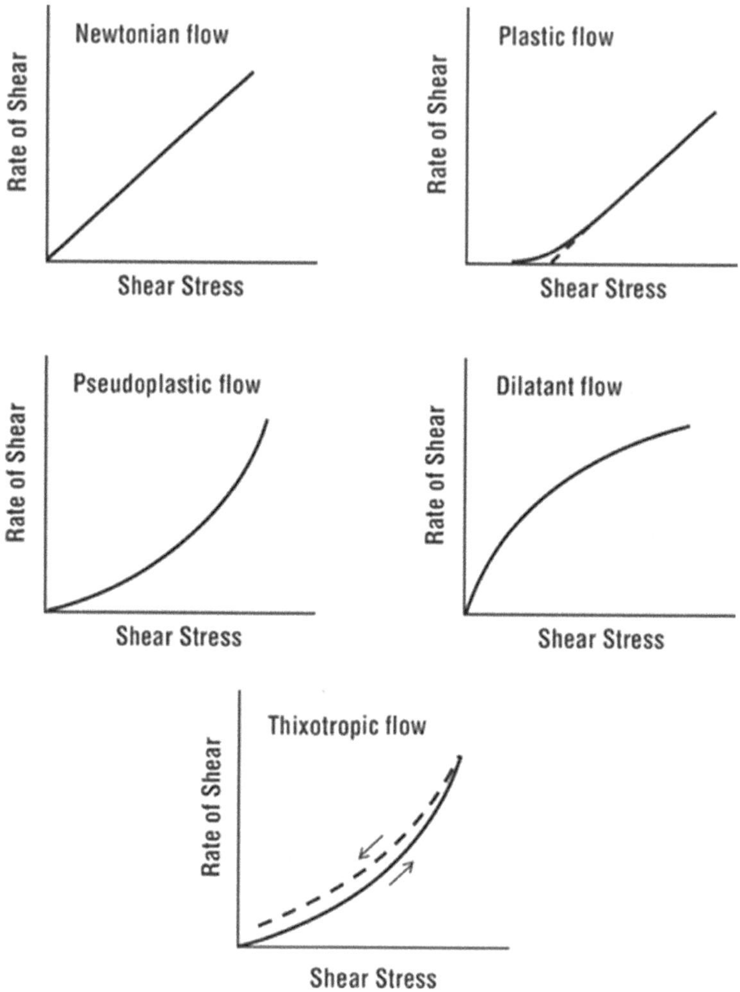

FIGURE 6.5 Rheograms of liquid dispersions. Adapted from (124).

The removal of shear stress produces a hysteresis in the rheograms of plastic and pseudoplastic systems, due to the breakdown of structure under shear stress. The restoration of structure occurs with time. This phenomenon is known as thixotropy. The rheogram of a thixotropic system depends upon the rate of applied and removed shear stress and the length of time the sample is subjected to the stress. Thixotropic suspensions are desirable for pharmaceutical formulations. These suspensions do not settle readily during storage, become fluid on shaking, remain in a suspended state as the dose is dispensed, and regain consistency for further storage (124).

BULK PARTICLE BEHAVIOR

This section discusses the correlation between particle size and the physical behavior of a powder, with particular emphasis on pharmaceutical processes. Pharmaceutical processes involving powders encompass a variety of dynamic effects ranging from powder flow and mixing, to dispersion as aerosols or liquid suspensions, to granulation and compression. The performance of the powders involved in pharmaceutical operations is highly dependent upon the particle size and particle size distribution of particles in the powder.

SELECTION OF THE APPROPRIATE PARTICLE SIZE EXPRESSION

It cannot be emphasized strongly enough that the particle size and statistics used to describe the powder should be chosen according to the pharmaceutical process in which the powder is to be used. A number of examples have been given relative to measures employed for the various intended uses. Although this concept is important for static effects, such as dissolution, it is even more critical with respect to particle dynamics. The dynamics of particles are highly dependent upon particle size; in many cases, the physics of the particle may vary with the square or cube of the particle diameter. For example, the terminal settling velocity of a particle in a viscous medium (air or water) increases with the square of the particle's geometric diameter. For particles intended for inhalation (i.e., less than 10 μm in size), the Cunningham slip correction factor must be added to the particle dynamics to account for the effect of ambient gas molecules with mean free path motion on the order of the size of the particles themselves. The Cunningham slip factor alters the settling velocity of particles by a factor of 1.02 for 10 μm particles, by 1.17 for 1 μm particles, and by 2.97 for 0.1 μm particles (125). Therefore, it is self-evident that particle size is of critical importance for inhalation powders, with differences of even a few μm significantly altering the dynamics of the particles. As can be seen from this example, no description of inhalation powders would be complete without stating the geometric diameter of the particles.

Although the geometric diameter of powder particles is important for most processes, other factors may be more influential in powder performance. For example, powder compressibility is determined primarily by particle volume, and dissolution is determined by particle surface area.

POWDER FLOW AND MIXING

The processing of most pharmaceutical products composed of powders will have a step that involves vibration. This step will often be a transfer process, such as die filling for tablet production or blister packing for aerosol unit dose production. Likewise, mixing processes are widely used in the preparation of pharmaceutical products. Tablet and dry powder aerosol dosage forms require mixing the active particles with excipients. Both of these processes are influenced by particle size. Milling, wherein particle size is reduced, is another example of a size-dependent process. Each of these processes are examined in more detail in the remainder of this chapter.

Vibration and Powder Transfer

Vertically vibrated powders have provoked significant interest among physicists. Vibration— in particular, sonic vibration—historically has been used in industrial powder processing. In a relevant industrial application, gas-fluidized beds were simultaneously vibrated using sonic energy generated by a loudspeaker (126, 127). The gas velocity required to fluidize the bed decreased when sonic energy was added to the system. The method was efficacious even in beds with particle sizes below 1 μm.

Physically vibrated powders can form heaps or complex patterns. The powder bed can be made to dilate if the vibration conditions are correct (128). For vertically vibrated powders, the amplitude of vibration is generally given by the function $z = a\cos(\omega t)$, where ω is the angular frequency in radians/second and a is the amplitude. For convection in the powder bed to occur, the vibrational acceleration must be larger than the gravitational acceleration. This condition can be written by using a dimensionless acceleration parameter. Dilation occurs when:

$$\Gamma = \frac{a\omega^2}{g} > 1 \tag{6.48}$$

where g is the gravitational constant. The dilation effect was found to be dependent upon the ambient gas. Dilation is significantly reduced if the ambient gas pressure drops below about 10 Torr (129). The postulate that interstitial gas is required for powder bed dilation was confirmed by experiments in a chamber where the interstitial gas was allowed to escape through holes in the side of the chamber as the powder was vibrated.

At vertical vibration accelerations with Γ greater than approximately 2.4, patterns appear on the surface of the granular material. Stripes, squares, or localized oscillations have been observed (129–133). A study by Duran demonstrated that powder beds tapped from below exhibit ripples and that particles can be ejected from the powder surface by tapping (134).

Vibration of powders causes segregation by size. Horizontally vibrated powders in closed containers exhibit a surprising phenomenon wherein large particles move to the bottom of the powder bed (135). This contrasts with the normally observed percolation that causes small particles to move to the bottom of the bed (136). This effect was determined to be due to upward convection near the walls of the vibrating container with a subsequent downward convection in the center.

Vibration can also result in consolidation of the powder bed. For consolidation, the degree of densification is a size-dependent phenomenon. Packing density varies with particle size when the powders have been sieved into discrete size ranges. Vertical vibration leads to denser packing of the powder bed than does horizontal vibration (137). Packing due to vibration is frequency dependent, but the frequency is largely particle size independent. These authors found that the optimal frequency of vibration for consolidation of lactose powders was between 100 and 200 Hz. These powders were divided into eight size ranges from around 20 to 150 μm.

Size segregation can also occur for nonvibratory powder transfer. In chute flow, fingers of powder appear across the advancing powder front due to the size segregation mechanism (138). In this situation, the coarser particles rise at the free surface due to the same percolation mechanism discussed previously. This leads to a recirculation zone at the front. The filling of a hopper leads to the segregation of particles by size, with smaller particles tending toward the center of the hopper (139, 140). Some remixing of particles can occur upon emptying of the hopper, but a net segregation is still observed. The degree of segregation depends on the manner in which the powder is discharged. The discharge flow pattern in turn depends upon the hopper geometry, orifice size, and wall roughness (141, 142). Hoppers or silos can empty by mass flow by which the entire content moves through the outlet or funnel flow in which only the center of the bulk powder flows through the outlet, leaving a dead zone (140).

The filling properties of capsules by a nozzle system can be similar to those by hopper flow. Specifically, the powder–wall interaction in the nozzle has been found to impact capsule filling in the same manner as for flow from a hopper (143). The particle size of the powder used in the fill is an important variable in capsule-filling properties (144, 145). When these methods of powder transfer are used, segregation mechanisms must be considered in the evaluation of end-product properties.

Pneumatic conveying may be used to move bulk solids over large distances. The process may reduce particle size due to the large amount of energy required to move the particles. In addition, segregation may occur due to particle size dependence in the deaeration behavior (146).

In general, particle size descriptors relevant to powder-handling processes are the geometric diameter and volume diameter. Both of these diameters determine the interparticulate forces and contact areas between powder particles. These diameters are also relevant to the bulk and tap densities, which, in turn, determine the settling properties that affect powder flow and transfer. For example, a direct correlation has been demonstrated between bulk density and capsule fill weight (147).

Mixing and Milling

The process of mixing has many applications in pharmaceutical processing. Much of the knowledge of the process is gained by experience (148), and the fundamental physics of the process is little understood. In addition, numerous types of mixers are used in pharmaceutical processing, from rotating vessels to vessels with moving baffles, blades, or screws (ribbon blenders). Each mixer will have different performance properties. Mixing cannot be viewed in isolation because the filling or emptying of the mixer can result in segregation, as discussed in the previous section.

A number of generic studies of the mixing of granular materials can be found in the physics literature. One example examined the mixing of granular materials in a tumbling mixer. In that study, Khakhar et al. developed a generic model for an arbitrarily shaped convex mixer (any two points in the mixer can be connected by a line that does not bisect the boundary of the mixer) (149). The model conditions required that the rotation rate be sufficient for particle transport to be continuous, rather than consisting of discrete avalanches, and that flow be constrained to an upper surface layer of limited depth. These conditions have been observed experimentally (150, 151). Under these conditions, a layer-by-layer map of particle trajectories was generated. For circular mixers, the trajectories were almost axially symmetric, and mixing between axial layers was poor. If the mixer geometry was elliptical or square, the trajectories became chaotic. When the model conditions were initiated with two segregated types of particles, the degree of segregation for a circular mixer decreased by only 10 percent after 20 rotations versus 35 percent for a square mixer. Khakhar et al. performed experimental simulations using 0.8 mm glass beads that provided results confirming the computational model. As they noted, particle sizes in their model were the same, whereas under practical mixing conditions, particles exist in a range of sizes. In this case, demixing by segregation of particle sizes would counter mixing by chaotic advection (149). The authors suggest that more complex blenders, such as V-blenders, probably perform by harnessing chaotic advection. By elucidating a model of mixing, Khakhar et al. provided a tool that could be employed in the design of efficient mixers (149).

A purely geometrical model has confirmed the timescale required for mixing in a circular drum (152). This model calculated the number of rotations required for the mixing of powder layers stacked vertically in the mixer. Here, mixing was defined as the occurrence of sufficient avalanches for the interface between powder layers to reach the surface through rotation. Uniform mixing, then, was not implied, but simply the disappearance of the initially pure surface regimes of the top powder. More than 10 revolutions were required for a drum 5 percent filled with greater than 25 percent of the volume occupied by the top powder. The mixing time decreased as the volume fraction of the top powder decreased below 25 percent and as the filling of the drum approached half full. For differently colored sodium chloride crystals that were otherwise identical particles, different authors found that the optimal circular mixer fill was 23 percent (153).

Mixing can induce size reduction by attrition in coincident milling. Alternatively, cohesive materials may aggregate during mixing due to surface tension forces that, in turn, are due to liquid bridging, mechanical interlocking, or electrostatic forces (154).

Ball milling is an example of a highly complex system for which knowledge of the relationships between the inputs (milling time, powder load, size and quantity of milling media, and rotation speed) and the output (milling rate, particle size, and distribution) is based on empirical observation. Caravati et al. modeled the motion of a single ball in a ball-milling device (155). Examination of the ball's motion, particularly the velocity of the ball before impact, could lead to predictions of the milling rate. The model was based upon an SPEX mixer/mill (model 8000, SPEX CertiPrep, Metuchen, N.J.). The motion of the vial and the frequency of collisions of the ball and vial using a piezoelectric shock sensor to detect impacts were used to validate the numerical simulation. The time series of impacts of the ball depended upon the powder charge in the mill. For a large powder charge, collisions of the ball and the wall of the mixer were plastic due to the cushioning effect of the powder between them. With plastic collisions, the impacts were highly periodic. As the powder charge was decreased, the collisions became more elastic. As a result, the ball motion lost its periodicity and eventually became aperiodic. A strange attractor was observed when the ball position and velocity were plotted in phase space for the plastic collisions. For this case, determination of the Lyapunov exponents confirmed that the motion of the ball was hyperchaotic. This study was the first step in the development of a means to predict milling properties from a physical model of the milling system (155). Correlation of the degree of chaos with the mill output parameters could be applied to optimization of input parameters for the desired output.

DISPERSION

Powders have been dispersed as aerosols for lung delivery since the invention of the dry powder inhaler (DPI) in the 1960s (156, 157). Interest in DPIs accelerated due to environmental concerns around the propellant in a pressurized metered dose inhaler (pMDI), the device that delivers the majority of respiratory drugs. Also, interest in delivering large molecules for systemic therapy was facilitated by DPIs (158). The discussion in this section is limited to particle size effects in therapeutic aerosol dispersion. Particle size is of critical importance because it affects not only dispersion but also deposition in the lungs. The range of particle sizes that can deposit in the lungs is limited to approximately 1–5 μm. Entire texts are available that discuss the physical pharmacy of aerosol dosage forms (159) or biological implications (160) of aerosol drug delivery.

At rest, granular materials act as plastic solids (161). In this state, the interparticle spacing is small, particle velocities are near zero, and the stresses on the powder bed are independent of velocity. The properties of the plastic regime determine the stability and slopes of heaps of material. Flow begins when a granular bed makes a transition to an inertial regime (162). Two other regimes of collective behavior exist. The fluidization regime is characterized by interparticle spacing on the order of the particle size. The interstitial fluid is the agent of momentum transfer and thus must overcome interparticle forces to initiate fluidization. Finally, the fourth regime is that of entrainment or suspension of the granular particles. Here the interparticle spacing is very large, and interaction between particles is negligible. The velocity of the particles is close to the velocity of the entrainment fluid. Generation of a powder aerosol involves the transition through the four states with the dynamics of each step being critically dependent upon the particle size. The conversion of a static powder bed to a powder aerosol requires dilation, flow, fluidization, and finally aerosol production (163, 164).

The efficiency of dispersion must also be considered in the design and formulation of dry powder aerosols. Dispersion requires the powder to overcome the interparticulate forces binding particles in bulk powder and to become entrained as single particles in the inhaled airstream. For respirable particles, these forces are dominated by the van der Waals force. All other factors being equal, the van der Waals force will decrease as the geometric particle size increases. Thus, in general terms, the probability of deposition in the deep lung acts in opposition to the efficiency of dispersion for DPIs (165).

The traditional approach to overcoming the difficulties in the dispersion of particles suitable for deposition in the lungs has been to blend these particles with larger carrier particles of lactose. The blended system optimizes interparticulate forces to create a mixture that will respond to the fluidization process. This system requires that the smaller particles be deaggregated from the larger carrier particles before they can enter the airways (165). In currently approved DPI devices, the forces required for deaggregation are provided by the patient, who generates shear forces in the DPI during inhalation (158). The blending of ternary mixtures has been shown to increase the proportion of respirable particles delivered by some DPIs (166). These blends comprise the micronized drug, the lactose carrier, and fine particles, typically lactose. The fine particles establish competition for binding sites and can result in a higher proportion of drug particles binding to weaker sites.

Engineering of low-density particles has gained popularity in recent years as an approach to overcoming the opposition between efficient dispersion and efficient aerosol delivery to the lungs (167, 168). This approach takes advantage of the relation between aerodynamic diameter, particle density, and geometric size. The balance of gravitational settling forces and the aerodynamic drag force determines this relationship. When these forces balance, the particle is said to have reached its terminal settling velocity. The aerodynamic diameter is defined as the diameter of a unit-density particle that has the same settling velocity as the non–unit-density particle being analyzed (169). Thus, the equation relating the geometric diameter to the aerodynamic diameter for spherical particles is given by:

$$d_a = \sqrt{\rho} d \tag{6.49}$$

where d_a is the aerodynamic diameter, ρ is the (non-unit) density of the particle, and d is the particle's geometric diameter. See Chapter 5, especially equations (5.4) and (5.5), for further details. Aerosols composed of particles with equal aerodynamic diameters but differing densities will necessarily have differing geometric diameters. Clearly, particles of less than unit density will have aerodynamic diameters that are smaller than their geometric diameters. Therefore, a particle with a geometric diameter >5 μm can be "respirable" provided the density is sufficiently small. This property, which allows large particles to penetrate deep into the lungs, is employed in the manufacture of large porous particles.

GRANULATION AND COMPRESSION

Granulation is the production of a powder by aggregation of smaller particles, typically used for the production of tablets and capsules. Bonding mechanisms employed during granulation are adhesion due to capillary pressure, adhesion due to liquid bridges, and solid bridges formed during drying (170). The liquid binding forces are responsible for the resultant particle size, whereas solid bridging is primarily responsible for granulate strength. Wet granulation involves the wetting of the seed particles with a liquid to create controlled conditions of agglomeration. The powder bed is kept in motion during this process either by flow through gas or other fluidization mechanisms such as vibration. Particle size effects during vibration were discussed in the previous section. The important point is that the energy input for fluidization must be increased during the process to overcome gravitational forces as the particles increase in size.

Compression of powders involves four stages: slippage and rearrangement of particles, elastic deformation of particles, plastic deformation or brittle fracture of particles, and compression of the solid crystal lattice (171). These mechanisms may not occur sequentially; for example, smaller particles formed by fragmentation may undergo another rearrangement step.

Tableting requires control of the particle size distribution of the components because this affects the final homogeneity of the active particles in the tablet and their bioavailability (172). In particular, matching the active and excipient particle sizes typically leads to better homogeneity due to a reduced tendency for particles to segregate. Much of this effect arises due to the flow properties of the powders during blending steps. The dissolution properties of the active drug depend upon the particle surface area, and a drug with poor aqueous solubility requires a larger surface area than a more soluble drug. In addition, greater tablet strength is often achieved in tablets formed of fine particles (173). Clearly, the determination of the particle surface area is important for powders used in tableting. Other problems arising from poor control of particle size distribution include variations in tablet weight and hardness. Variations in flow properties due to particle size can result in poor flow from hoppers and bins or flooding of tablet machines.

CONCLUSION

Characterization of fundamental particle properties is the foundation of establishing specifications that support the quality and performance of products in which they are incorporated. Particle size is generally described in terms of equivalent spherical diameters. Particle shape may be characterized by using geometric shape factors or mathematical approaches, such as Fourier and fractal analyses. Surface roughness is generally determined by profilometry and may be characterized by using parameters such as rugosity, root mean square deviation, and maximum peak-to-valley height. The surface area of particles is most commonly determined by the BET nitrogen adsorption method. The bulk density of powders depends on the true density and packing arrangement of particles, which in turn is affected by the particle size distribution, shape, and adhesional properties.

Particle behavior is examined with a particular focus on particle adhesion and motion within bulk powders, gaseous dispersions, and liquid dispersions. Electrostatic charge measurements are performed using the Faraday well or cage for dry powders, or electrophoresis for liquid dispersion systems. Adhesion force techniques measure the detachment of either single or multiple particles. Single-particle detachment is measured by using microbalance, pendulum, or atomic force microscopy methods. Multiple-particle detachment is measured by using centrifugation, vibration, impaction, and aerodynamic techniques. Fundamental particle properties, such as chemical composition, particle size, particle shape, and surface roughness, greatly affect particle adhesion. Environmental conditions, including storage duration and humidity, also affect particle adhesion. The blending of powders may involve the random mixing of noninteracting particles or may utilize the adhesion of particles, known as interactive mixing.

Powder flow may be characterized by using static bed measurements, such as the angle of repose or spatula, Carr's compressibility index, the Hausner ratio, or shear cell measurements, or dynamic flow measurements, including the rotating drum or vibrating spatula. Interparticulate forces need to be overcome for powder flow to commence, so factors affecting powder flow are similar to those affecting particle adhesion. The gas fluidization of powders is classified according to Geldart (1986). Particle motion in gaseous and liquid dispersions involves either diffusion by Brownian motion or gravitational sedimentation. Stokes' law describes the rate of particle settling in laminar conditions under gravity. Slip correction factors are applied to Stokes' law for very small particles. A variation of Stokes' law enables the calculation of the rate of particle settling in a centrifugal field.

The rheological behavior of liquid dispersions is determined from the rate of shear as a function of shearing stress. The flow behavior is described as Newtonian and non-Newtonian, which includes plastic, pseudoplastic, and dilatant behavior. Plastic and pseudoplastic systems may exhibit thixotropy.

Fundamental particle properties have a great influence on particle behavior. An adequate and specific characterization of pharmaceutical particulate systems is required to understand particle behavior.

This chapter ties the specific utilization of pharmaceutical powders to the appropriate measure of particle size. The physical behavior of a powder particle is closely correlated to at least one measure of its size or characteristics. One or two of these characteristics will frequently be more predictive of its performance than others in a given pharmaceutical process. However, the choice of characteristic highly depends on the desired process.

REFERENCES

1. Allen T. Particle size, shape and distribution. In: Allen T, editor. *Particle Size Measurement*. London, UK: Chapman and Hall; 1975. p. 74–112.
2. Yee BM, Sinko PJ. Pharmaceutical Micromeritics. In: Sinko PJ, editor. *Martin's Physical Pharmacy*. Philadelphia, PA: Wolters Kluwer; 2023. p. 528–50.
3. Dallavalle JM. Shape and size distribution of particles. *Micromeritics*. New York, NY: Pitman; 1948. p. 41–67.
4. Hesketh HE. Introduction. *Fine Particles in Gaseous Media*. Chelsea, UK: Lewis Publishers; 1986. p. 1–22.
5. Podczeck F, Newton M, James MB. Variations in the adhesion force between a drug and carrier particles as a result of changes in relative humidity. *Int J Pharm*. 1997;149:151–60.
6. Mandelbrot B. *The Fractal Geometry of Nature*. New York, NY: WH Freeman and Co; 1982.
7. Mandelbrot B. How long is the coast of Britain? Statistical self-similarity and fractional dimension. *Science*. 1967;156:636–8.

8. Richardson LF. The problem of contiguity: An appendix to statistics of deadly quarrels. *General Systems Yearbook*. 1961;6:139–97.
9. Kaye B. Fractal description of fine particle systems. In: Beddow JK, editor. *Particle Characterization in Technology*. 1. Applications and Microanalysis. Boca Raton, FL: CRC Press; 1984. p. 81–100.
10. Thibert R, Akbarieh M, Tawashi R. Application of fractal dimension in the study of surface ruggedness of granular solids and excipients. *J Pharm Sci*. 1988;77:724–6.
11. Fernandez-Hervas MJ, Holgado MA, Raboasco AM, Fini A. Use of fractal geometry on the characterization of particle morphology: Application to the diclofenac hydroxyethylpirrolidine salt. *Int J Pharm*. 1994;108:187–94.
12. Fini A, Fernandez-Hervas MJ, Holgado MA, Rodriguez L, Cavalleri C, Passerini N, et al. Fractal analysis of β-cyclodextrin-indomethacin particles compared to ultrasound. *J Pharm Sci*. 1997;86:1303–9.
13. Luerkens DW, Beddow JK, Vetter AF. Theory of morphological analysis. In: Beddow JK, editor. *Particle Characterization in Technology*. 2. Morphological Analysis. Boca Raton, FL: CRC Press; 1984. p. 3–14.
14. Hundal HS, Rohani S, Wood HC, Pons MN. Particle shape characterization using image analysis and neural networks. *Powder Tech*. 1997;91:217–27.
15. Hickey AJ, Concessio NM. Descriptors of irregular particle morphology and powder properties. *Adv Drug Deliv Rev*. 1997;26:29–40.
16. Podczeck F. Fundamentals of adhesion of particles to surfaces. *Particle-particle adhesion in pharmaceutical powder handling*. London, UK: Imperial College Press; 1998. p. 1–80.
17. Shakesheff KM, Davies MC, Roberts CJ, Tendler SJB, Williams PM. The role of scanning probe microscopy in drug delivery research. *Crit Rev Ther Drug Carrier Sys*. 1996;13:225–56.
18. Avnir D, Jaroniec M. An isotherm equation for adsorption on fractal surfaces of heterogeneous porous materials. *Langmuir*. 1989;5:1431–3.
19. Dallavalle JM. Determination of particle surface. *Micromeritics*. New York, NY: Pitman; 1948. p. 327–42.
20. Dallavalle JM. Introduction. *Micromeritics*. New York, NY: Pitman; 1948. p. 3–13.
21. Dallavalle JM. Characteristics of packings. *Micromeritics*. New York, NY: Pitman; 1948. p. 123–48.
22. Stewart PJ. Particle interactions in pharmaceutical systems. *Pharmacy Int*. 1986;7:146–9.
23. Zimon AD. Adhesion in a gas medium: Sources of adhesion. In: *Adhesion of Dust and Powder*. New York, NY: Consultant Bureau; 1982. p. 93–144.
24. Sinko PJ. Diffusion. In: Sinko PJ, editor. *Martin's Physical Pharmacy and Pharmaceutical Sciences*. 8th ed. Philadelphia, PA: Wolters-Kluwer; 2023. p. 281–317.
25. Lobenberg B, Silva DA. Disintegration, dissolution and drug release. In: Sinko PJ, editor. *Martin's Physical Pharmacy and Pharmaceutical Sciences*. 8th ed. Philadelphia, PA: Wolters-Kluwer; 2023. p. 318–40.
26. Visser J. Particle adhesion and removal: A review. *Part Sci Tech*. 1995;13:169–96.
27. Staniforth JN, Rees JE. Electrostatic charge interactions in ordered powder mixes. *J Pharmacy Pharmacol*. 1982;34:69–76.
28. Kulvanich P, Stewart PJ. An evaluation of the air stream Faraday cage in the electrostatic charge measurement of interactive drug systems. *Int J Pharm*. 1987;36:243–52.
29. Byron PR, Peart J, Staniforth JN. Aerosol electrostatics I: Properties of fine powders before and after aerosolization by dry powder inhalers. *Pharm Res*. 1997;14:698–705.
30. Keskinan J, Pietarinen K, Lehtimaki M. Electrical low pressure impactor. *J Aerosol Sci*. 1992;23:353–60.

31. Rhodes CT. Disperse systems. In: Bamker GS, Rhodes CT, editors. *Modern Pharmaceutics*. New York, NY: Marcel Dekker; 1989. p. 327–54.
32. Visser J. Particle adhesion and removal: A review. *Particulate Sci Tech*. 1995;13:169–96.
33. Visser J. An invited review. Van der Waals and other cohesive forces affecting powder fluidization. *Powder Tech*. 1989;58:1–10.
34. Corn M. The adhesion of solid particles to solid surfaces II. *J Air Poll Control Assoc*. 1961;11:523–8.
35. Morgan BB. The adhesion and cohesion of fine particles. *Br Coal Util Res Assoc Monthly Bull*. 1961;25:125–37.
36. Israelachvili JN. Adhesion. In: *Intermolecular Surface Forces*. London: Academic Press; 1991. p. 312–37.
37. Padmadisastra Y, Kennedy RA, Stewart PJ. Influence of carrier moisture adsorption capacity on the degree of adhesion of interactive mixtures. *Int J Pharm*. 1994;104:R1–R4.
38. Zimon AD. Methods for determining adhesive force. In: *Adhesion of Dust and Powder*. New York, NY: Consultants Bureau; 1982. p. 69–91.
39. Israelachvili JN. Contrast between intermolecular, interparticle and intersurface forces. In: *Intermolecular and Surface Forces*. New York, NY: Academic Press; 1991. p. 152–75.
40. Corn M. Adhesion of particles. In: Davies CN, editor. *Aerosol Science*. New York, NY: Academic Press; 1966. p. 359–92.
41. Corn M. The adhesion of solid particles to solid surfaces II. *J Air Poll Control Assoc*. 1961;11:566–75.
42. Binnig G, Quate CF, Gerber C. Atomic force microscopy. *Phys Rev Lett*. 1986;56:930–3.
43. Heinz WF, Hoh JH. Spatially resolved force spectroscopy of biological surfaces using atomic force microscopy. *Nanotechnology*. 1999;17:143–50.
44. Ducker WA, Senden TJ, Pashley RM. Direct measurement of colloidal forces using atomic force microscopy. *Nature*. 1991;353:239–41.
45. Claesson PM, Ederth T, Bergeron V, Rutland MW. Techniques for measuring surface forces. *Adv Colloid Interface Sci*. 1996;67:119–83.
46. Mizes HA. Surface roughness and particle adhesion. *J Adhesion*. 1995;51:155–65.
47. Fuji M, Machida K, Takei T, Watanabe T, Chizazawa M. Effect of wettability on adhesion force between silica particles evaluated by atomic force microscopy measurement as a function of relative humidity. *Langmuir*. 1999;15:4584–9.
48. Sindel U, Zimmerman I. Direct measurement of interparticulate forces in particulate solids using atomic force microscopy. Proceedings of World Congress on Particle Technology. 1998. p. 31chemE.
49. Louey MD, Mulvaney P, Stewart PJ. Characterization of adhesional properties of lactose carriers using atomic force microscopy. *J Pharm Biomed Anal*. 2001;25:559–67.
50. Kulvanich P, Stewart PJ. The effect of particle size and concentration on the adhesive characteristics of a model drug-carrier interactive system. *J Pharmacy Pharmacol*. 1987;39:673–8.
51. Kulvanich P, Stewart PJ. Fundamental considerations in the measurement of adhesional forces between particles using the centrifuge method. *Int J Pharm*. 1987;35:111–20.
52. Kordecki MC, Orr C. Adhesion of solid particles to solid surfaces. *Am Med Assoc Arch Environ Health*. 1960;1:13–21.
53. Boehme G, Krupp H, Rabenhorst H, Sandstede G. Adhesion measurements involving small particles. *Trans Inst Chem Engrs*. 1962;40:252–9.
54. Podczeck F, Newton JM, James MB. Assessment of adhesion and autoadhesion phenomena between particles and surfaces. Part I: The investigation of autoadhesion phenomena of salmeterol xinafoate and lactose monohydrate particles using compacted powder surfaces. *J Adhesion Sci Tech*. 1994;8:1459–72.

55. Staniforth JN, Rees JE, Lai FK, Hersey JA. Interparticulate forces in binary and ternary ordered powder mixes. *J Pharmacy Pharmacol.* 1982;34:141–5.
56. Staniforth JN, Rees JE, Lai FK, Hersey JA. Determination of interparticulate forces in ordered powder mixes. *J Pharmacy Pharmacol.* 1981;33:485–90.
57. Booth SW, Newton JM. Experimental investigation of adhesion between powders and surfaces. *J Pharmacy Pharmacol.* 1987;39:679–84.
58. Podczeck F, Newton JM, James MB. The adhesion strength of particles of salmeterol base and a series of salmeterol salts to compacted lactose monohydrate surfaces. *J Adhesion Sci Tech.* 1995;9:1547–58.
59. Okada J, Matsuda Y, Yoshinobu F. Measurement of the adhesive force of pharmaceutical powders by the centrifuge method. *Yakugaku Zasshi.* 1969;89:1539–44.
60. Podczeck F, Newton JM, James MB. Adhesion and autoadhesion measurements of micronized particles of pharmaceutical powders to compacted powder surfaces. *Chem Pharm Bull.* 1995;43:1953–7.
61. Laycock S, Staniforth JN. Problems encountered in accurate determination of interparticulate forces in ordered mixes. *Part Sci Tech.* 1983;1:259–68.
62. Laycock S, Staniforth JN. A method for determining interparticulate force in ordered mixes. *Labo-Pharma:Problems Tech.* 1984;32:185–9.
63. Podczeck F, Newton JM, James MB. Assessment of adhesion and autoadhesion forces between particles and surfaces. Part II: The investigation of adhesion phenomena of salmeterol xinafoate and lactose monohydrate particles in particle-on-particle and particle-on-surface contact. *J Adhesion Sci Tech.* 1995;9:475–86.
64. Podczeck F, Newton JM. Development of an ultracentrifuge technique to determine adhesion and friction properties between particles and surfaces. *J Pharm Sci.* 1995;84:1067–71.
65. Neumann, BS. The flow properties of powders. In: Bean HS, Carless, JE and Beckett, AH, editors. *Advances in pharmaceutical sciences*, New York: Academic Press; pp. 181–221.
66. Yip CW, Hersey JA. Ordered or random mixing: Choice of system and mixer. *Drug Dev Ind Pharm.* 1977;3:429–38.
67. Bryan L, Rungvejhavuttivittaya Y, Stewart PJ. Mixing and demixing of microdose quantities of sodium salicylate in a direct compression vehicle. *Powder Tech.* 1979;22:147–51.
68. Stewart PJ. Influence of magnesium stearate on the homogeneity of a prednisolone-granule ordered mix. *Drug Dev Ind Pharm.* 1981;7:485–95.
69. Staniforth JN, Rees JE. Effect of vibration time, frequency and acceleration on drug content uniformity. *J Pharmacy Pharmacol.* 1982;34:700–6.
70. Staniforth JN, Rees JE. Segregation of vibrated powder mixes containing different concentrations of fine potassium chloride and tablet excipients. *J Pharmacy Pharmacol.* 1983;35:549–54.
71. Staniforth JN, Ahmed HA, Lockwood PJ. Quality assurance in pharmaceutical powder processing: Current developments. *Drug Dev Ind Pharm.* 1989;15:909–26.
72. Zimon AD. Detachment of adherent particles in an air stream. In: *Adhesion of Dust and Powder.* New York, NY: Consultants Bureau; 1982. p. 307–47.
73. Mullins ME, Michaels LP, Menon V, Locke B, Ranade MB. Effect of geometry on particle adhesion. *Aerosol Sci Tech.* 1992;17:105–18.
74. Otsuka A, Iida K, Danjo K, Sunada H. Measurements of the adhesive force between particles of powdered organic substances and a glass substrate by means of impact separation method. I: Effect of temperature. *Chem Pharm Bull.* 1983;31:4483–8.

75. Concessio NM, vanOort MM, Hickey AJ. Impact force separation measurements: Their relevance in powder aerosol formulation. In: Dalby RN, Byron PR, Farr SJ, editors. *Respiratory Drug Delivery IV.* Buffalo Grove, IL: Interpharm Press; 1998. p. 251–8.
76. Otsuka A, Iida K, Danjo K, Sunada H. Measurements of the adhesive force between particles of powdered materials and a glass substrate by means of the impact force separation method. III: Effect of particle shape and surface asperity. *Chem Pharm Bull.* 1988;36:741–9.
77. Concessio NM, vanOort MM, Hickey AJ. Pharmaceutical dry powder aerosols. Correlation of powder properties with dose delivery and implications for pharmacodynamic effect. *Pharm Res.* 1999;16:828–34.
78. Zimon AD. Fundamental concepts of particle adhesion. In: *Adhesion of Dust and Powder.* New York, NY: Consultants Bureau; 1982. p. 1–30.
79. Nystrom C, Malmqvist K. Studies on direct compression of tablets. I. The effect of particle size in mixing finely divided powders with granules. *Acta Pharm Suecica.* 1980;17:282–7.
80. Zimon AD, Volkova TS. Effect of surface roughness on dust adhesion. *Colloid J USSR (English Translation).* 1965;27:306–7.
81. Zimon AD. Adhesion of variously shaped particles to rough surfaces. In: *Adhesion of Dust and Powder.* New York, NY: Consultants Bureau; 1982. p. 145–72.
82. Staniforth JN. Order out of chaos. *J Pharmacy Pharmacol.* 1987;39:329–34.
83. Staniforth JN. Powder flow. In: Aulton ME, editor. *Pharmaceutics: The Science of Dosage Form Design.* New York, NY: Churchill Livingstone; 1988. p. 600–15.
84. Ziskind G, Fichman M, Gutfinger C. Resuspension of particulates from surfaces to turbulent flows-review and analysis. *J Aerosol Sci.* 1995;26:613–44.
85. Ward GH, Schultze RK. Process-induced crystallinity changes in albuterol sulfate and its effect on powder physical stability. *Pharm Res.* 1995;12:773–9.
86. Kulvanich P, Stewart PJ. The effect of blending time on particle adhesion in model interactive system. *J Pharmacy Pharmacol.* 1987;39:732–3.
87. Lam KK, Newton JM. Investigation of applied compression on the adhesion to a substrate surface. *Powder Tech.* 1991;65:167–75.
88. Lam KK, Newton JM. The influence of the time of application on contact pressure on particle adhesion to a substrate surface. *Powder Tech.* 1993;76:149–54.
89. Lam KK, Newton JM. Effect of temperature on particulate solid adhesion to substrate surface. *Powder Tech.* 1992;73:267–74.
90. Lam KK, Newton JM. Influence of particle size on the adhesion behavior of powders, after application of an initial press on force. *Powder Tech.* 1992;73:117–25.
91. Kulvanich P, Stewart PJ. Correlation between total adhesion and charge decay of a model interactive system during storage. *Int J Pharm.* 1987;39:51–7.
92. Kulvanich P, Stewart PJ. Influence of relative humidity on the adhesive properties of a model interactive system. *J Pharmacy Pharmacol.* 1988;40:453–8.
93. Zimon AD. Adhesion of solid particles to a plane surface 2: Influence of air humidity on adhesion. *Colloid J USSR (English Translation).* 1963;25:265–8.
94. Podczeck F, Newton JM, James MB. The influence of constant and changing relative humidity of the air on the autoadhesion force between pharmaceutical powder particles. *Int J Pharm.* 1996;145:221–9.
95. Padmadisastra Y, Kennedy RA, Stewart PJ. Solid bridge formation in sulphonamide-emdex interactive systems. *Int J Pharm.* 1994;112:55–63.
96. Crowder TM, Hickey AJ. The physics of powder flow applied to pharmaceutical solids. *Pharm Tech.* 2000;24:50–8.

97. Neumann BS. The flow properties of powders. In: Bean HS, Carless JE, Beckett AH, editors. *Advances in Pharmaceutical Sciences.* New York, NY: Academic Press; 1967. p. 181–221.
98. Carr RR. Evaluating flow properties of solids. *Chem Eng.* 1965;18 (Jan):163–8.
99. Jenike AW. Flow of solids in bulk handling systems. *Bull Univ Utah.* 1954;64.
100. Amidon GE. Physical and mechanical property characterization of powders. In: Brittain HG, editor. *Physical Characterization of Pharmaceutical Solids.* New York, NY: Marcel Dekker; 1995. p. 281–319.
101. Smith DLO, Lohnes RA. Behavior of bulk solids. In: Beddow JK, editor. *Particle Characterization in Technology.* 1. Applications and Microanalysis. Boca Raton, FL: CRC Press; 1984. p. 102–33.
102. Crowder TM, Sethuraman V, Fields TB, Hickey AJ. Signal processing and analysis applied to powder behavior in a rotating drum. *Part Part Sys Char.* 1999;16:191–6.
103. Hickey AJ, Concessio NM. Descriptors of irregular particle morphology and powder properties. *Adv Drug Deliv Rev.* 1997;26:29–40.
104. Kaye B. Characterizing the flowability of a powder using the concepts of fractal geometry and chaos theory. *Part Part Sys Char.* 1997;14:53–66.
105. Hickey AJ, Concessio NM. Flow properties of selected pharmaceutical powders from a vibrating spatula. *Part Part Sys Char.* 1994;11:457–62.
106. Crowder TM, Hickey AJ. An instrument for rapid powder flow measurement and temporal fractal analysis. *Part Part Sys Char.* 1999;16:32–4.
107. Rietema K. General introduction. In: *The Dynamics of Fine Powders.* New York, NY: Elsevier; 1991. p. 1–18.
108. Geldart D. Characterization of fluidized powders. In: *Gas Fluidization Technology.* New York, NY: Wiley; 1986. p. 33–51.
109. Lantz Jr. RJ, Schwartz JB. Mixing. In: Lieberman HA, Lachman L, editors. *Pharmaceutical Dosage Forms.* New York, NY: Marcel Dekker; 1981. p. 1–53.
110. Hersey JA. Ordered mixing: A new concept in powder mixing practice. *Powder Tech.* 1975;11:41–4.
111. Hersey JA. The development and applicability of powder mixing theory. *Int J Pharm Tech Prod Manufac.* 1979;1:6–13.
112. Carstensen JT. Two-component systems. In: *Solid Pharmaceutics: Mechanical Properties and Rate Phenomena.* New York, NY: Academic Press; 1980. p. 102–34.
113. Dallavalle JM. Diffusion of particles. In: *Micromeritics.* New York, NY: Pitman; 1948. p. 164–79.
114. Hesketh HE. Unidirectional motion of particles. In: *Fine Particles in Gaseous Media.* Chelsea, UK: Lewis Publishers; 1986. p. 23–44.
115. Hinds WC, Zhu Y. Properties of gases. In: *Aerosol Technology: Properties, Behavior and Measurement of Airborne Particles.* 3rd ed. New York, NY: Wiley; 2022. p. 15–36.
116. Hinds WC, Zhu Y. *Aerosol Technology: Properties, Behavior and Measurement of Airborne Particles.* 3rd ed. New York, NY: Wiley; 2022.
117. Hinds WC, Zhu Y. Uniform particle motion. In: *Aerosol Technology: Properties, Behavior, and Measurement of Airborne Particles.* 3rd ed. New York, NY: Wiley; 2022. p. 37–64.
118. Carmichael GR. The effect of shape on particle solids flow. In: Beddow JK, editor. *Particle Characterization in Technology.* 2. Morphological Analysis. Boca Raton, FL: CRC Press; 1984. p. 205–21.
119. Hinds WC, Zhu Y. Appendix A12. In: *Aerosol Technology: Properties, Behavior and Measurement of Airborne Particles.* New York, NY: Wiley; 2022. p. 402.
120. Dallavalle JM. Collection and separation of particulate matter from air. In: *Micromeritics.* New York, NY: Pitman; 1948.

121. Hinds WC, Zhu Y. Straight line acceleration and curvilinear particle motion. In: *Aerosol Technology: Properties, Behavior and Measurement of Airborne Particles.* 3rd ed. New York, NY: Wiley; 2022. p. 97–120.
122. Sinko PJ. Coarse dispersions. In: Sinko PJ, editor. *Martin's Physical Pharmacy and Pharmaceutical Sciences.* 8th ed. Philadelphia, PA: Wolters-Kluwer; 2023. p. 551–76.
123. Giddings CJ. Field-flow fractionation: Analysis of macromolecular, colloidal and particulate materials. *Science.* 1993;260:1456–65.
124. Sinko PJ. Rheology. In: *Martin's Physical Pharmacy and Pharmaceutical Sciences.* 8th ed. Philadelphia, PA: Wolters-Kluwer; 2023. p. 260–80.
125. Crowder TM, Rosati JA, Schroeter JD, Hickey AJ, Martonen TB. Fundamental effects of particle morphology on lung delivery: Predictions of Stokes' law and the particular relevance to dry powder inhaler formulation. *Pharm Res.* 2002;19:239–45.
126. Leu L-P, Li JT, Chen CM. Fluidization of group B particles in an acoustic field. *Powder Tech.* 1997;94:23–8.
127. Mori S, Haruta T, Yamamoto A, Yamada I, Mizutani E. Vibrofluidization of very fine particles. *Int Chem Engr.* 1991;31:475–80.
128. VanDoorn E, Behringer RP. Dilation of a vibrated granular layer. *Europhysics Lett.* 1997;40:387–92.
129. Pak HK, VanDoorn E, Behringer RP. Effects of ambient gases on granular materials under vertical vibration. *Phys Rev Lett.* 1995;74:4643–6.
130. Jaeger HM, Nagel SR, Behringer RP. The physics of granular materials. *Physics Today.* 1996;49:32–8.
131. Pak HK, Behringer RP. Surface waves in vertically vibrated granular materials. *Phys Rev Lett.* 1993;71:1832–5.
132. Umbanhower PB, Melo F, Swinney HL. Localized excitations in a vertically vibrated granular layer. *Nature.* 1996;382:793–6.
133. Melo F, Umbanhowar PB, Swinney HL. Hexagons, kinks and disorder in oscillated granular layers. *Phys Rev Lett.* 1995;75:3838–41.
134. Duran J. Ripples in tapped or blown powder. *Phys Rev Lett.* 2000;84:5126–9.
135. Liffman K, Metcalfe G, Cleary P. Convection due to horizontal shaking. In: Behringer RP, Jenkins JT, editors. *Powders and Grains.* Rotterdam, The Netherlands: Balkema; 1997. p. 405–8.
136. Barker G, Grimson M. The physics of muesli. *New Scientist.* 1990;126:37–40.
137. Woodhead PJ, Chapman SR, Newton JM. The vibratory consolidation of particle size fractions of powders. *J Pharmacy Pharmacol.* 1983;35:621–6.
138. Pouliquen O. Fingering instability in granular flows. In: Behringer RP, Jenkins JT, editors. *Powders and Grains.* Rotterdam, The Netherlands: Balkema; 1997. p. 451–4.
139. Standish N. Studies of size segregation in filling and emptying a hopper. *Powder Tech.* 1985;45:43–56.
140. Schulze D. Storage, feeding, proportioning. In: Chulia D, Deleuil D, Pourcelot Y, editors. *Powder Technology and Pharmaceutical Processes.* Amsterdam, The Netherlands: Elsevier; 1994. p. 285–317.
141. Kafui KD, Thornton C. Some observations on granular flow in hoppers and silos. In: Behringer RP, Jenkins JT, editors. *Powders and Grains.* Rotterdam, The Netherlands: Balkema; 1997. p. 511–4.
142. Khati S, Shahrour I. An experimental study of the evolution of density during flow of granular materials. In: Behringer RP, Jenkins JT, editors. *Powders and Grains.* Rotterdam, The Netherlands: Balkema; 1997. p. 491–4.
143. Jolliffe JG, Newton JM, Walters JK. Theoretical considerations of the filling of pharmaceutical hard gelatin capsules. *Powder Tech.* 1980;27:189–95.

144. Jolliffe IG, Newton JM. An investigation of the relationship between particle size and compression during capsule filling with instrumented mG2 simulator. *J Pharmacy Pharmacol*. 1982;34:415–9.
145. Hogan J, Shue PI, Podczeck F, Newton JM. Investigations into the relationship between drug properties, filling, and the release of drugs from hard gelatin capsules using multivariate statistical analysis. *Pharm Res*. 1996;13:944–9.
146. Hilgraf P. Pneumatic conveying. In: Chulia D, Deleuil M, Pourcelot Y, editors. *Powder Technology and Pharmaceutical Processes*. Amsterdam, The Netherlands: Elsevier; 1994.
147. Newton JM, Bader F. The prediction of the bulk densities of powder mixtures, and its relationship to the filling of hard gelatin capsules. *J Pharmacy Pharmacol*. 1981;33:621–6.
148. Poux M, Fayolle P, Bertrand J, Bridoux D, Bousquet J. Powder mixing - some practical rules applied to agitated systems. *Powder Tech*. 1991;68:213–34.
149. Khakhar DV, McCarthy JJ, Gilchrist JF, Ottino JM. Chaotic mixing of granular materials in two dimensional tumbling mixers. *Chaos*. 1999;9:195–205.
150. Henein H, Brimacombe JK, Watkinson AP. Experimental study of transverse bed motion in rotary kilns. *Metallurg Trans B*. 1983;14:191–7.
151. Nakagawa M, Altobelli SA, Caprihan A, Fukushima E. NMR measurement and approximate derivation of the velocity depth profile of granular flow in a rotating partially filled, horizontal cylinder. In: Behringer RP, Jenkins JT, editors. *Powders and Grains*. Rotterdam, The Netherlands: Balkema; 1997. p. 447–50.
152. Dorogovtsev SN. Avalanche mixing of granular materials: Time taken for the disappearance of the pure fraction. *Tech Phys*. 1997;42:1346–8.
153. Metcalfe G, Shinbrot T, McCarthy JJ, Ottino JM. Avalanche mixing of granular solids. *Nature*. 1995;374:39–41.
154. Bridgwater J. Mixing. In: Chulia D, Deleuil M, Poucelot Y, editors. *Powder Technology and Pharmaceutical Processes*. Amsterdam, The Netherlands: Elsevier; 1994. p. 347–57.
155. Caravati C, Delogu F, Cocco G, Rustici M. Hyperchaotic qualities of the ball motion in ball milling device. *Chaos*. 1999;9:219–26.
156. Clark AR. Medical aerosol inhalers. Past, present and future. *Aerosol Sci Tech*. 1995;22:374–91.
157. Hickey AJ, Dunbar CA. A new millennium for inhaler technology. *Pharm Tech*. 1997;21:116–25.
158. Crowder TM, Louey MD, Sethuraman VV, Smyth HDC, Hickey AJ. 2001: An odyssey in inhaler formulations and design. *Pharm Tech*. 2001;25:99–113.
159. Hickey AJ, daRocha SP. *Pharmaceutical Inhalation Aerosol Technology*. 3rd ed. Boca Raton, FL: CRC Press; 2019.
160. Hickey AJ, Mansour HM. *Inhalation Aerosols: Physical, and Biological Basis for Therapy*. 3rd ed. Boca Raton, FL: CRC Press; 2019.
161. Castellanos A, Valverde JM, Perez AT, Ramos A, Watkins PK. Flow regimes in fine cohesive powders. *Phys Rev Lett*. 1999;82:1156–9.
162. Evesque P. Analysis of the statistics of sandpile avalanches using soil-mechanics results and concepts. *Phys Rev A*. 1991;43:2720–40.
163. Dunbar CA, Hickey AJ, Holzner P. Dispersion and characterization of pharmaceutical dry powder inhalers. *KONA Powder and Particle*. 1998;16:7–45.
164. Telko MJ, Hickey AJ. Dry powder inhaler formulation. *Resp Care*. 2004;50:1209–27.
165. Hickey AJ, Concessio NM, van Oort NN, Platz RM. Factors influencing the dispersion of dry powders as aerosols. *Pharm Tech*. 1994;18:58–64.

166. Zeng XM, Pandal KH, Martin GP. The influence of lactose carrier on the content homogeneity and dispersibility of beclomethasone dipropionate from dry powder aerosols. *Int J Pharm.* 2000;197:41–52.
167. Edwards DA, Ben-Jebria A, Langer R. Recent advances in pulmonary drug delivery using large, porous inhaled particles. *J Appl Physiol.* 1998;85:379–85.
168. Edwards DA, Hanes J, Caponetti G, Hrkach J, Ben-Jebria A, Eskew ML, et al. Large porous particles for pulmonary drug delivery. *Science.* 1997;276:1868–71.
169. Reist PC. *Aerosol Science and Technology.* New York: McGraw-Hill; 1993.
170. Ormos ZD. Granulation and coating. In: Chulia D, Deleuil M, Pourcelot Y, editors. *Powder Technology and Pharmaceutical Processes.* Amsterdam, The Netherlands: Elsevier; 1994. p. 359–76.
171. Doelker E. Assessment of powder compaction. In: Chulia D, Deleuil M, Pourcelot Y, editors. *Powder Technology and Pharmaceutical Processes.* Amsterdam, The Netherlands: Elsevier; 1994. p. 403–71.
172. Lewis A, Simpkin G. Tabletting-an industrial viewpoint. In: Chulia D, Deleuil M, Pourcelot Y, editors. *Powder Technology and Pharmaceutical Processes.* Amsterdam, The Netherlands: Elsevier; 1994. p. 473–92.
173. Rudnic EM, Kottke MK. Tablet dosage forms. In: Banker G, Rhodes CT, editors. *Modern Pharmaceutics.* New York, NY: Marcel Dekker; 1996. p. 333–94.

7 Particulate Pharmaceutical Processes and Products

MANUFACTURING PROCESSES AFFECTED BY PARTICLE COMPOSITION AND BEHAVIOR

The ability to manufacture dosage forms containing particles is dictated by the ability to move bulk powder to the site of individual dosage form construction, for example, capsule fillers or tablet dies. Bulk powder movement requires flow properties that are unimpeded by individual particle behavior. It is essential that a heterogeneous powder composed of drug and additive particles remain uniformly dispersed during transport to assure the accuracy and reproducibility of drug content in each of the dosage forms. Consequently, tendencies to segregate either on the basis of composition, size, shape, density, or some other property of the blend should be addressed during product development.

The following sections consider: (1) the composition of the powder that is employed to prepare the dosage form, (2) the processes that are influenced by powder behavior, and (3) the sequence of events that determine the appearance of the drug in the molecular state from the dosage form.

PARTICLE COMPONENTS OF DOSAGE FORM

Particle manufacture was described in Chapter 3. There are a wide variety of methods but the most commonly employed for the preparation of drugs and excipients are milling and spray drying(1–3). Drug particles are characterized as described in Chapter 5 to ensure their quality and performance as components of the final dosage form. Similar scrutiny is applied to additives as drug, since their function in the formulation will depend on the control of their properties and dictate the accuracy and reproducibility of the drug delivery (4). Where additives in the particulate form are employed their starting properties, for example, particle size distribution, must be controlled and their compatibility with the drug ensured (4, 5).

Additives are employed in particle-based dosage forms for a variety of reasons and in different forms (6). Historically the simplest way in which additives can be employed in the preparation of the dosage form is by blending particles of the additive with particles of drug to achieve a homogeneous distribution of particles (7). The intent of preparing the blend is that each of the components can confer the desired

DOI: 10.1201/9781003470151-7

benefit to the performance of the formulation. The primary objective is to deliver the dose of drug accurately and reproducibly to assure efficacy and safety during product development. Consequently, assuring the uniformity of drug in each dose is central to the objective of effective delivery (8).

The additives employed are frequently used to minimize the effect of processing on the performance of the dosage form. Silicon dioxide may be added to aid the consistency of flow during the movement of the powder (9, 10). Sugars, such as lactose, may be added as bulking agents for very low-dose drugs to allow ease of dispensing from the bulk (11, 12). Other additives may be employed to control the performance of the dosage once it has been prepared. For example, starch encourages swelling, disintegration, and deaggregation to present drug particles at the site of absorption for dissolution (13). Salts and surfactants may be employed to aid in pH control and dissolution (14–17).

Processing Steps in Formulation and Dosage Form Manufacture

The use of particles and powders in dosage forms may require additional processing steps. These processing steps are usually intended to facilitate production-scale manufacturing to ensure the quality, performance, safety, and efficacy of the final dosage form. It is rarely the case that drug particles alone are employed in the formulation. The presence of additives may be required depending on the processing, stability, and biopharmaceutical requirements as outlined above.

The four major processing steps for oral solid dose forms that must be considered are blending, granulation, filling, and compression (18, 19). These are staged in a sequence, as illustrated in Figure 7.1, that allows sampling (Chapter 4) for batch processes, or in-line or online analysis for continuous processes, allowing characterization and analysis (Chapter 5) to allow control that assures the quality of the product (20–23).

Blending

To establish the quality of the product in terms of dose uniformity and performance, a method of combining the component particles to achieve homogeneous distribution in the powder must be developed. Blending is a well-established method that achieves this objective. The factors influencing blend uniformity include particle properties, such as moisture content, morphology, size distribution and density, and blending conditions, including energy input (speed of rotation or agitation in V-blenders, ribbon blenders, and oscillatory blenders) and blend time (7, 18, 24). While the numerical values differ as a function of the variables described, a general exponential decay in variability (relative standard deviation) is observed with the time of blending. Process optimization is employed to reduce the time and energy required to achieve a homogeneous blend (25, 26).

Granulation

When manufacturing on a production scale, one way to minimize deviation in dose uniformity is to add an additional layer of control to the formulation by preparing

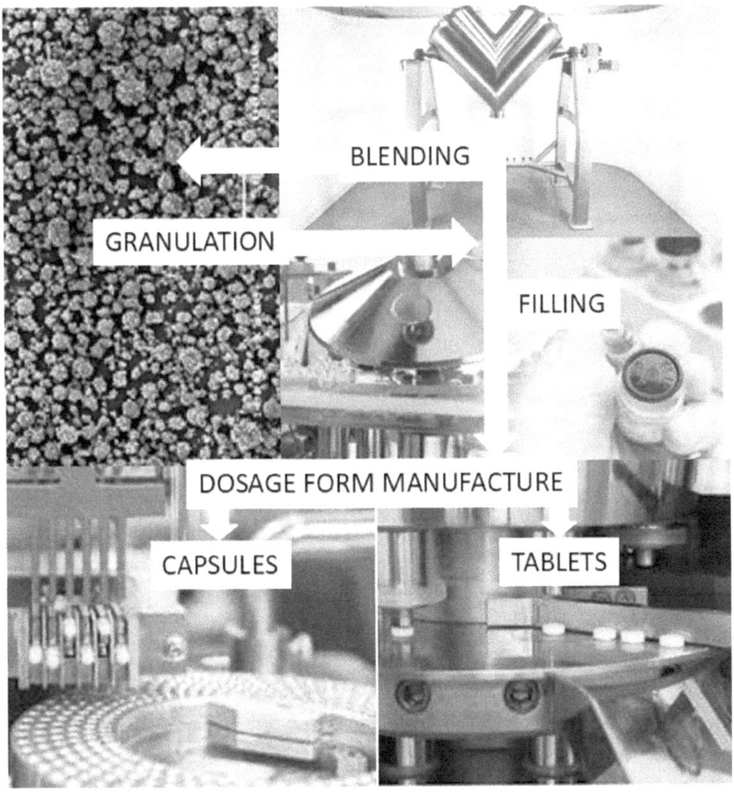

FIGURE 7.1 Powder processing steps in pharmaceutical solid dosage form manufacturing.

granules (large aggregates) of the component particles and using these to assist in subsequent manipulation of the powder (27). Granulation can be achieved by wet and dry processes (28, 29). The wet process employs a binding agent in solution to create a slurry of dispersed particles that upon drying contain a homogeneous dispersion of drug particles. The dry process involves passing a homogeneous powder blend between rollers that apply a compression force. The compressed powder emerges from the rollers in granules of large aggregates.

Granulated particles are substantially larger than the component powder and are consequently easier to manipulate in terms of flow and filling required for solid oral dosage forms. Dry powder inhaler products do not use granulation processes because the primary particle size of the drug in the formulation dictates the delivery to the lungs as an aerosol (30, 31).

Filling

To prepare individual dosage forms at the desired dose for efficacy and safety, the powder containing all ingredients in a homogeneous dispersion must be conveyed from a bulk hopper to the site of the dosage form manufacture, i.e., the die in a tablet press or a capsule in a capsule filler (32, 33). The flow properties of the powder are important to the ease with which the powder can be moved. The properties that are measured to assess the potential of the powder to be suitable for a filling operation can be divided into static and dynamic methods. Static methods include angle of repose, bulk, and tapped density with derived metrics of Carr's index and Hausner ratio (33–35). In addition, specific tools have been developed that measure dynamic flow through orifices, for example, Flodex, which is of direct relevance to the movement of powders during filling processes (36, 37). Vibrational flow methods have also been employed to evaluate the energy required to move a powder where a relatively free-flowing powder requires a small stimulus to move (38, 39). Finally, the Jenike shear cell reflects the movement of powder in a constrained space (40).

Batch and Continuous Processing

Batch processing requires considerable monitoring and control to ensure the quality and performance of the product. Since the physicochemical properties of the formulation and the conditions for processing (blending, granulation, filling) are fixed at the beginning of a batch process the method of control involves establishing the quality of the starting materials, setting the operating specifications of equipment, and implementing rigorous analytical standards to ensure the product meets specifications. This approach follows the principles of Quality by Design and Six Sigma methodology (41, 42). The approach requires formulation and process, design, measurement, analysis, improvement, and control to establish quality and performance. An unfortunate aspect of batch manufacturing is that there is little opportunity for in-process control. Consequently, in the event of an oversight in measurement and analysis or equipment failure the process cannot be corrected in real time or stopped before the designated sample analysis indicates that the product does not meet specification. However, the control of input variables dictated by Quality by Design principles minimizes the prospects of failure to meet specification, increases efficiency, and reduces the cost associated with poor control (41, 43). In addition, discrete element analysis can be employed to model batch processes and establish optimal operating conditions (44). Continuous processing is an attractive alternative to batch processing because real-time controls can be established with in-line and on-line analysis of critical quality attributes, and equipment operating conditions can be adjusted as required to minimize product failure in production. The additional controls available in continuous processing serve to eliminate any remaining risk associated with acting in an untimely manner associated with batch processing (22, 23, 45).

Product and Intermediate Characterization

Product and intermediate characterization is intended to assure the quality and performance within specifications that have been shown to correlate with safety and efficacy. These specifications are dictated by regulatory guidances, such as those issued by the US FDA, or published as public standards through national compendia such as the United States Pharmacopeia.

For particle and powder products, the physicochemical properties must be characterized as described in previous chapters to establish the crystallinity, crystal habit, polymorphism, moisture content, particle morphology, and particle size distribution. In addition, certain performance characteristics such as dissolution for oral dosage forms or aerodynamic particle size distribution for aerosol products should be measured (46).

These properties may be defined as critical to product quality and performance in which case they would be studied on transport and storage to ensure the stability of the drug product (47, 48). The intermediates may have been studied for their criticality during development and additives selected to control conditions such as pH or moisture content in the final dosage form.

The ability with continuous processing to monitor drug uniformity, particle size distribution, and homogeneity at various points in the process by in-line or online methods gives a layer of control that is independent of the operator and subject to automation which renders the process resource, time, and cost efficient.

FINAL DOSAGE FORM

The final dosage form may consist of free-flowing powder or a compressed solid composed of individual particles. In the case of free-flowing powder, it is important to the performance of the dosage form that the particles are presented individually for whatever subsequent phenomena occur, leading to the bioavailability of the drug. Any factors that might impact dispersion should be considered including, moisture content and electrostatic charge (49, 50). Unique dosage forms such as nasal and inhalation products also require the aerodynamic behavior to be assessed as this may impact respiratory tract deposition and delivered dose (51, 52). The overall stability of the powder must also be evaluated.

PROPERTIES AFFECTING DOSAGE FORM PERFORMANCE

The physicochemical properties of particulates that play a role in dosage form performance may be divided into those that impact on (a) static particle-particle interactions, (b) dynamic particle–particle interactions, and (c) particle surrounding medium interactions.

Moisture

The uptake of moisture by particles and powders gives rise to both chemical and physical instability. The process of moisture uptake can be driven by a number of phenomena. Hygroscopic drug and excipient particles spontaneously take up water from the atmosphere (53). Where the major component of the particle exhibits polymorphism including amorphous regions, the propensity to take up water may differ from the stable crystalline form of the drug or excipient (54). Variations in temperature and relative humidity may contribute to moisture uptake or loss, either of which can change the surface properties and potentially the overall structure of the particles leading to behavioral differences (55–57). In powders that have the potential to take up water, the interstitial space can also be occupied by water, giving rise to suction potential due to capillary forces that bind the particles into aggregates (58, 59).

Ideally, particles are prepared in the most physically stable polymorph and any processes or environmental conditions that may give rise to sources of instability are controlled to limit chemical degradation or physical changes such as Ostwald ripening, recrystallization, liquid crystal formation, and aggregation.

Electrostatic Charge

In dry environments, particles often become electrostatically charged (60–62). Triboelectric charges at particle surfaces are also induced by contact with surrounding materials, such as processing equipment and containers (63, 64). This can result in aggregation and interaction with adjacent surfaces affecting dose reproducibility; it can also cause processing challenges in filling and sealing of capsules, blisters, or other containment units (65).

Aerodynamic Behavior

Drug particles employed in pharmaceutical inhalation aerosols are expected to be dispersed into aerodynamic particle sizes suitable for nasal or pulmonary delivery (51, 52). Aerodynamic particle size distribution dictates deposition and disposition from the respiratory tract which in turn correlates with safety and efficacy (66). Particles >10 μm deposit efficiently in the nose. Particles >10 μm do not penetrate to the lungs. Indeed, particles >5 μm have limited potential to enter the lungs. Consequently, as a guide to aerosol products intended for drug delivery to the lungs incorporate particles <5 μm prepared by a variety of methods (67). Chapter 5 provides a discussion of the aerodynamic behavior of particles relevant to inhalation delivery.

STABILITY

The stability of a dosage form may be dictated by the properties of the particles it contains, potentially impacting its quality and performance. The critical factors influencing the stability of the particles are polymorphism, hygroscopicity, additive incompatibility, presence of microbes, presence of foreign particulates or impurities, and sensitivity to light and/or heat (68, 69). The nature of the instability can be at the molecular level leading to chemical degradation of the drug due to interactions with chemical or energetic sources of instability. At the physical level, the proximity of particles during processing may lead to changes in morphology, surface properties, or extent of interaction leading to aggregate or agglomerate formation.

Investigation of chemical stability follows the prescribed protocols employing either extremes of temperature to probe limits (Arrhenius equation) of stability of particles alone or standard conditions of temperature and humidity to evaluate long-term storage stability of the dosage form (70).

DRUG RELEASE

Drug release from the dosage form consists of several steps depending on the way in which the particles have been combined. In solid oral dosage forms where compacts of powders composed of a variety of particles have been prepared, there are notably three steps: disintegration of the single dosage form, deaggregation of the integral particles, and dissolution of the particles into the component molecules which are subsequently transported in the surrounding liquid medium or body fluids (Figure 7.2).

Disintegration and Deaggregation

Disintegration and deaggregation are phenomena that occur concurrently with dissolution in solid oral dosage forms. Experimentally these are captured by conducting dissolution experiments. The most commonly conducted dissolution methods are the USP basket and paddle methods which control the dimension of the dissolution vessel, the volume and composition of the dissolution medium in which the dosage form is immersed, and the rotation speed of either the basket containing the dosage form or the paddle above the dosage form (71). Controlling the key variables in the experimental apparatus allows accurate and reproducible data to be collected from which the performance of the dosage form can be specified. This is useful in characterizing the quality and performance of novel dosage forms with respect to safety and efficacy and set performance criteria for in vitro bioequivalence of generic products.

Dissolution

The rate of drug dissolution can be described by the Noyes–Whitney equation (72):

$$\frac{dc}{dt} = \frac{kDS(C_s - C_t)}{Vh} \tag{7.1}$$

where dc/dt is the rate of dissolution, k is the dissolution constant, D is the coefficient of diffusion, S is the surface area, V is the volume of the dissolution media, h is the thickness of the stagnant diffusion layer, C_s is the drug concentration at the interface, and C_t is the drug concentration in the bulk media.

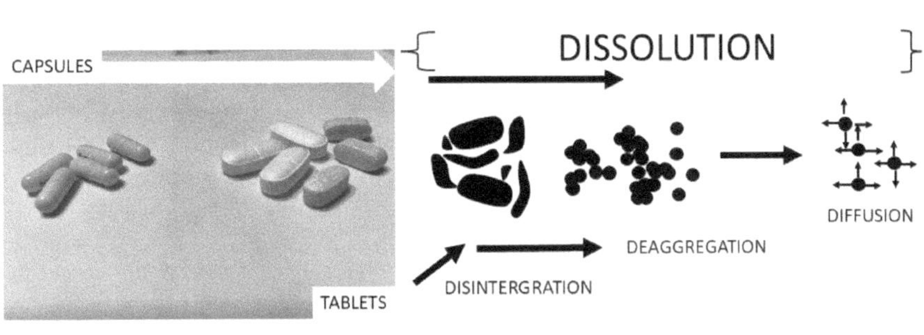

FIGURE 7.2 Phenomena that contribute to the dissolution of solid-state dosage forms, capsules, and tablets, composed of particulates.

Once the drug has dissolved, dispersion into the bulk medium, or transfer from the surface of the particle, and subsequently through barriers such as cell membranes is dictated by a number of phenomena. At the surface of the particle and through membranes where no mixing occurs, diffusion is the dominant mechanism of transport. Where mixing occurs, forces move bulk fluid carrying dissolved drug rapidly to other regions.

Diffusion

The interface of a particle with the surrounding liquid medium in an in vitro dissolution apparatus or at the in vivo site of adsorption is restricted from movement. The thickness of this unstirred layer depends on the environment in which the particle is located. Where there is no mixing, the layer is thick; while in a mixed environment, the outer regions of this layer adjacent to the bulk liquid are disturbed and the unstirred layer is then thinner than in the unmixed condition. The general expression for Fick's first law is as follows:

$$J = -D\left(\frac{dC}{dx}\right) \tag{7.2}$$

Where J is the diffusion flux (mass of drug per unit area per time); D is the diffusion coefficient, C is the concentration of drug and x is the distance traveled. This is expression is sufficient to predict the diffusion of drug in the boundary layers at specific distances from the particle surface. However, we are frequently asked to consider movement through not a hypothetical layer (distance) but through a formal barrier such as a membrane.

The rate of drug diffusion across an unstirred layer or a membrane can also be described by Fick's first law of diffusion (73, 74):

$$\frac{dQ}{dt} = DAR\left(\frac{C_g - C_b}{X}\right) \tag{7.3}$$

where dQ/dt is the rate of diffusion (or flux), D is the diffusion coefficient of the drug, A is the surface area of the membrane available for drug diffusion, R is the partition coefficient of the drug between the membrane and the aqueous fluid, C_g is the drug concentration at the site of absorption, C_b is the drug concentration in the blood, and X is the thickness of the membrane. The concentration gradient (C_g - C_b) is the driving force for diffusion across the membrane. Under sink conditions, in which the drug concentration at the site of absorption is much greater than the drug concentration in the blood, the equation may be simplified to a first-order kinetic process (75):

$$\frac{dQ}{dt} = KC_g \tag{7.4}$$

Under physiological conditions, sink conditions apply due to large blood volume compared with the volume of gut fluid, and rapid clearance of the drug from the site of absorption by systemic circulation.

Passive diffusion mechanisms produce a linear relationship between the amount of the drug absorbed and the dose administered. In contrast, active transport mechanisms may become saturated and generally produce a curvilinear relationship that reaches a plateau (75).

Since diffusion is occurring throughout the disintegration and deaggregation of a tablet, the mathematical description is confounded by the change in surface area that occurs as the process is initiated. This was first recognized in the Higuchi equation that indicates the rate of dissolution following a square root relationship between concentration and time following initiation of dissolution (76).

SUMMARY

In previous chapters, the manufacture, properties, and characterization of particles were considered. This exposition highlighted the control over the physicochemical nature of these finely divided solids that can be achieved and the extent to which monitoring and control can be used to assure quality. In this chapter, the way in which particles may be combined in complex systems to achieve the objective of therapeutic drug delivery is outlined. It is important to recognize that the route of administration requires different performance characteristics of the dosage form. At the simplest level this might require free-flowing powders, controlled aggregates or granules, or compressed single solids in the final state. The processes in which the initial powder formulation plays a significant role are blending, granulation, filling, and compression. The properties and composition of the particles and the processing conditions may be monitored and controlled to ensure the quality of the final product. Its performance must then be established in terms of critical properties, notably dissolution or aerodynamic behavior. In vitro methods are available to assess these performance attributes that then support regulatory review (Chapter 8) and in vivo assessment of safety and efficacy (Chapter 9).

REFERENCES

1. Naik S, Chaudhuri B. Quantifying dry milling in pharmaceutical processing: A review on experimental and modeling approaches. *J Pharm Sci*. 2015;104:2401–13.
2. Ordoubadi M, Wang H, Vehring R. Mechanistic formulation design of spray dried powders. *KONA Powder Part J*. 2023;40:149–71.
3. Parrot EL. Milling of pharmaceutical solids. *J Pharm Sci*. 1974;63:813–29.
4. Elder DP, Kuentz M, Holm R. Pharmaceutical excipients - quality, regulatory and biopharmaceutical considerations. *Eur J Pharm Sci*. 2016;87:88–99.
5. RC, Bhandari S. Drug-excipient compatibility screening - role of thermoanalytical and spectroscopic techniques. *J Pharm Biomed Anal*. 2014;87:82–97.
6. Pockle RD, Masareddy RS, Patil AS, Patil PD. A comprehensive review on pharmaceutical excipients. *Ther Deliv*. 2023;14:443–58.
7. Pernenkil L, Cooney CL. A review on the continuous blending of powders. *Chem Eng Sci*. 2006;61:720–42.
8. Asachi M, Nourafkan E, Hassanpour A. A review of current techniques for the evaluation of powder mixing. *Adv Powder Technol*. 2018;29:1525–49.

9. Blanco D, Antikainen O, Raikkonen H, Yliruusi Y, Juppo AM. Effect of colloidal silicon dioxide and moisture on powder flow properties: Predicting in-process performance using image-based analysis. *Int J Pharm.* 2021;597:120344.
10. Jonat S, Hasenzahi S, Gray A, Schmidt PC. Mechanism of glidants: Investigation of the effect of colloidal silicon dioxide types on powder flow by atomic force and scanning electron microscopy. *J Pharm Sci.* 2004;93:2635–44.
11. Shi C, Zhao H, Fang Y, Shen L, Zhao L. Lactose in tablets: Functionality, critical material attributes, applications, modifications and co-processed excipients. *Drug Discov Today.* 2023;28:103696.
12. Hebbink GA, Dickhoff BHJ. Application of lactose in the pharmaceutical industry. In: Poques M, Lindner C, editors. *Lactose - Evolutionary Role, Health Effects, and Applications.* New York, Academic Press; 2019. p. 175–229.
13. Desai PM, Liew CV, Heng PWS. Review of disintegrants and the disintegration phenomena. *J Pharm Sci.* 2016;105:2545–55.
14. Gupta D, Bhatia D, Dave V, Sutariya V, Gupta SV. Salts of therapeutic agents: Chemical, physicochemical and biological considerations. *Molecules.* 2018;23:1719.
15. He Y, Orton E, Yang D. The selection of a pharmaceutical salt-The effect of the acidity of the counterion on its solubility and potential biopharmaceutical performance. *J Pharm Sci.* 2018;107:419–25.
16. vanderMerwe J, Steenekamp J, Steyn D, Hamman J. The role of functional excipients in solid oral dosage forms to overcome poor drug dissolution and bioavailability. *Pharmaceutics.* 2020;12:393.
17. Correa-Soto CE, Gao Y, Indulkar AS, Zhang GGZ, Taylor LS. Role of surfactants in improving release from higher drug loading amorphous solid dispersions. *Int J Pharm.* 2022;625:122120.
18. Hickey AJ. *Pharmaceutical Process Engineering.* 3rd ed. Boca Raton, FL: CRC Press; 2025.
19. Jindal AB. *Pharmaceutical Process Engineering and Scale-up Principles.* New York, NY: Springer; 2023.
20. Rantanen J, Khinast J. The future of pharmaceutical manufacturing sciences. *J Pharm Sci.* 2015;104:3612–38.
21. Sacher S, Pams J, Rehri J, Khinast JG. PAT implementation for advanced process control in solid dosage manufacturing - A practical guide. *Int J Pharm.* 2022;613:121408.
22. Narang AS, Dubey A. *Continuous Pharmaceutical Processing and Process Analytical Technology.* Boca Raton, FL: CRC Press; 2023.
23. Muzzio FJ, Goodrich CL, Alexander A, Arratia P, Yang H, Sudah O, et al. Sampling and characterization of pharmaceutical powders and granular blends. *Int J Pharm.* 2003;250:51–64.
24. Anderson CA, Velez NL. Blending and characterization of pharmaceutical powders. In: Merkus HG, Meesters GMH, Oostra W, editors. *Particles Nanoparticles Pharmaceut Products.* New York, NY: Springer; 2018.
25. Yeom SB, Choi DH. Scale-up strategy in quality by design approach for pharmaceutical blending process with discrete element analysis. *Pharmaceutics.* 2019;11:264.
26. Jaspers M, Kulkarni SS, Tegel F, Roelofs TP, deWit MTW, Janssen PHM, et al. Batch versus continuous blending of binary and ternary pharmaceutical powder mixtures. *Int J Pharm X.* 2022;4:100111.
27. Parikh DM. *Handbook of Pharmaceutical Granulation.* New York, NY: Informa Healthcare; 2010.
28. Faure A, York P, Rowe RC. Process control and scale-up of pharmaceutical wet granulation processes: A review. *Eur J Pharm Biopharm.* 2001;52:269–77.

29. Kleinebudde P. Roll compaction/dry granulation: Pharmaceutical applications. *Eur J Pharm Biopharm.* 2004;58:317–26.
30. Telko MJ, Hickey AJ. Dry powder inhaler formulation. *Respir Care.* 2005;50:1209–27.
31. Hickey AJ. Dry powder inhalers: An overview. *J Aerosol Med Pulm Drug Deliv.* 2023;36:29104.
32. Finke JH, Kwade A. Powder processing in pharmaceutical applications - In-depth understanding and modeling. *Pharmaceutics.* 2021;13:128.
33. Shah RB, Tawakkul MA, Khan MA. Comparative evaluation of flow for pharmaceutical powders and granules. *AAPS PharmSciTech.* 2008;9:250–8.
34. Wong AC-Y. Use of angle of repose and bulk densities for powder characterization and the prediction of minimum fluidization and minimum bubbling velocities. *Chem Eng Sci.* 2002;57:2635–40.
35. Hao T. Understanding empirical powder flowability criteria scaled by Hausner ratio or Carr index with the analogous viscosity concept. *RSC Adv.* 2015;5:57212–5.
36. Zhou X, Nauka E, Narang A, Mao C. Flow function of pharmaceutical powders at low-stress conditions can be inferred using a simple flow-through orifice device. *J Pharm Sci.* 2020;109:2009–17.
37. Taylor MK, Ginsberg J, Hickey AJ, Gheya F. Composite method to quantify powder flow as a screening method in early tablet and capsule formulation. *AAPS PharmSciTech.* 2000;1:18.
38. AJH, Concessio NM. Flow properties of selected pharmaceutical powders from a vibrating spatula. *Part Part Syst Char.* 1994;11:457–62.
39. Dunst P, Bornmann P, Hemsel T, Sextro W. Vibration-assisted handling of dry fine powders. *Actuators.* 2018;7:18.
40. Jager PD. Assessment of pharmaceutical powder flowability using shear cell-based methods and application of Jenike's methodology. *J Pharm Sci.* 2015;104:3804–13.
41. Yu LX, Amidon G, Khan MA, Hoag SW, Polii J, Raju GK, et al. Understanding pharmaceutical quality by design. *AAPS J.* 2014;16:771–83.
42. Byrne B, McDermott O, Noonan J. Applying lean six sigma methodology to a pharmaceutical manufacturing facility: A case study. *Processes.* 2021;9:550.
43. Yu LX. Pharmaceutical quality by design: Product and process development, understanding and control. *Pharm Res.* 2008;25:781–91.
44. Ketterhagen WR, TamEnde M, Hancock BC. Process modeling in the pharmaceutical industry using the discrete element method. *J Pharm Sci.* 2009;98:442–70.
45. Rossi CV. A comparative investment analysis of batch versus continuous pharmaceutical manufacturing technologies. *J Pharm Innov.* 2022;17:1373–91.
46. Tougas TP, Mitchell JP. *Good Cascade Impactor Practices, AIM and EDA for Orally Inhaled Products.* New York, NY: Springer; 2013.
47. Jurgens K. Influence of particle size and shape of drug product - aspects of formulation and performance. *Am Pharm Rev.* 2006;9:61–5.
48. Azad MA, Capellades G, Wang AB, Klee DM, Hammersmith G, Rapp K, et al. Impact of critical material attributes (CMAs)-Particle shape on miniature pharmaceutical unit operations. *AAPS PharmSciTech.* 2021;22:98.
49. Ahlneck C, Zografi G. The molecular basis of moisture effects on the physical and chemical stability of drugs in the solid state. *Int J Pharm.* 1990;62:87–95.
50. Karner S, Urbanetz NA. The impact of electrostatic charge in pharmaceutical powders with specific focus on inhalation powders. *J Aerosol Sci.* 2011;42:428–45.
51. Hickey AJ, daRocha SR. *Pharmaceutical Inhalation Aerosol Technology.* 3rd ed. Boca Raton, FL: CRC Press; 2019.
52. Kassinos S, Backman P, Conway J, Hickey AJ. *Inhaled Medicines.* New York, NY: Elsevier; 2021.

53. Newman AW, Reutzel-Edens SM, Zografi G. Characterization of the "hygroscopic" properties of active pharmaceutical ingredients. *J Pharm Sci.* 2008;97:1047–59.
54. Muller T, Krehl R, Schiewe J, Weiler C, Steckel H. Influence of small amorphous amounts of hydrophilic and hydrophobic APIs on storage stability of dry powder inhalation products. *Eur J Pharm and Biopharm.* 2015;92:130–8.
55. Karde V, Ghorai C. Influence of surface modification on wettability and surface energy characteristics of pharmaceutical excipient powders. *Int J Pharm.* 2014;475:351–63.
56. Crouter A, Briens L. The effect of moisture on flowability of pharmaceutical excipients. *AAPS PharmSciTech.* 2013;15:65–74.
57. Koumbogle K, Gosselin R, Gitzhofer F, Abatzoglou N. Moisture behavior of pharmaceutical powder during the tableting process. *Pharmaceutics.* 2023;15:1652.
58. Butt H-J. Capillary forces: Influence of roughness and heterogeneity. *Langmuir.* 2008;24:4715–21.
59. Yang L, Sega M, Harteng J. Capillary-bridge forces between solid particles: Insights from lattice Boltzmann simulations. *AICHE J.* 2021;67:e17350.
60. Wong J, Kwok PCL, Chan H-K. Electrostatics in pharmaceutical solids. *Chem Eng Sci.* 2015;125:225–37.
61. Gruberova L, Kratochvil B, Jegorov A. Effect of electrostatic charge on the production and composition of pharmaceuticals. *Chemicke Listy.* 2014;108:853–8.
62. Kaialy W. A review of factors affecting electrostatic charging of pharmaceuticals and adhesive mixtures for inhalation. *Int J Pharm.* 2016;503:262–76.
63. Ali M, Ghadiri M. Analysis of triboelectric charging of particles due to aerodynamic dispersion by a pulse of pressurized air jet. *Adv Powder Tech.* 2017;28:2735–40.
64. Ghori MU, Conway BR. Triboelectrification of pharmaceutical powders: A critical review. *Brit J Pharmacy.* 2018;3:550.
65. Desai PM, Tan BMJ, Liew CV, Chan LW, Heng PWS. Impact of electrostatics on processing and product performance of pharmaceutical solids. *Curr Pharm Des.* 2015;21:5923–9.
66. Hickey AJ. Preclinical and clinical considerations. *Inhaled Pharmaceutical Product Development Perspectives.* New York, NY: Elsevier; 2018. p. 53–71.
67. Hickey AJ. Lung deposition and clearance of pharmaceutical aerosols: What can be learned from inhalation toxicology and industrial hygiene. *Aerosol Sci Tech.* 1993 (online 2007);18:290–304.
68. Carstensen J, Rhodes C. *Drug Stability, Revised, and Expanded.* Boca Raton, FL: CRC Press; 2007.
69. Loftsson T. *Drug Stability for Pharmaceutical Scientists.* New York, NY: Academic Press; 2014.
70. Gonzalez-Gonzalez O, Ramirez IO, Ramirez BI, O'Connell P, Bellesteros MP, Torrado JJ, et al. Drug stability: ICH versus accelerated predictive stability studies. *Pharmaceutics.* 2022;14:2324.
71. USP. *General Chapter <711> Dissolution. United States Pharmacopeia.* Rockville, MD: USP; 2025.
72. Dokoumetzidis A, Macheros P. A century of dissolution research: From Noyes Whitney to the Biopharmaceutics Classification System. *Int J Pharm.* 2006;321:1–11.
73. Bird RB, Stewart WE, Lightfoot EN. *Transport Phenomena.* 2nd ed. 2007.
74. Li X, Jasti B. Theory and applications of diffusion and dissolution. In: Ghosh TK, Jasti BR, editors. *Theory and Practice of Contemporary Pharmaceutics.* Boca Raton, FL: CRC Press; 2004. p. 197–216.
75. Mayersohn M. Principles of drug absorption. In: Florence AT, Siepmann J, editors. *Modern Pharmaceutics.* 1st & 5th ed. Boca Raton, FL: CRC Press; 2010. p. 23–80.
76. Siepmann J, Peppas NA. Higuchi equation: Derivation, applications, use and misuse. *Int J Pharm.* 2011;418:6–12.

8 Regulatory Considerations for Pharmaceutical Development

The overall stages of pharmaceutical development include drug discovery, pre-clinical research, clinical development, regulatory submission, and commercial production (Figure 8.1). During the drug discovery stage, the biological target is identified and validated, high-throughput screening identifies "hits" (molecules that show activity against the target), and the chemical structure of "hits" is optimized through structure-activity relationship studies to improve the potency, selectivity, and bioavailability of lead compounds. During the pre-clinical stage, in vitro and in vivo studies are performed in cell cultures or animal models to evaluate the efficacy, safety, and pharmacokinetics of the lead candidates. Adsorption, distribution,

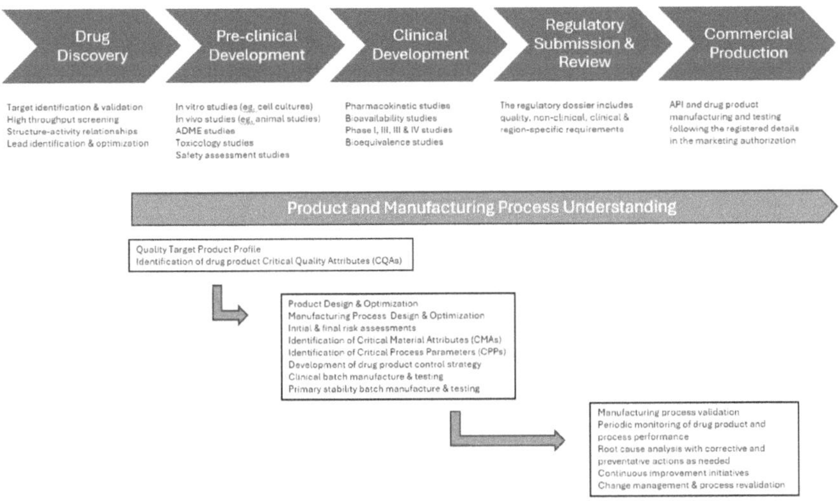

FIGURE 8.1 Stages of pharmaceutical development.

DOI: 10.1201/9781003470151-8

metabolism, and excretion (ADME) properties are evaluated. Toxicology and safety assessment studies are required to support regulatory submissions. These can include genotoxicity, carcinogenicity, and reproductive toxicology studies. The International Council for Harmonization (ICH) Safety Guidelines (1) describe specific requirements for safety studies.

During clinical development, clinical studies are conducted on human subjects to assess the drug product's pharmacokinetics, safety, efficacy, and tolerability. Phase I studies are generally conducted in healthy subjects to evaluate the drug candidate's safety, tolerability, pharmacokinetics, and pharmacodynamics. Phase II studies are generally conducted in the patient population to assess safety and efficacy and to identify the dosages and frequency for future studies. Phase III studies are pivotal studies conducted in patient populations to demonstrate the drug candidate's safety, efficacy, and tolerability. The pivotal study is designed with adequate statistical power to demonstrate the investigational product's non-inferiority, superiority, or equivalence against the existing standard of care. The International Council for Harmonization (ICH) Efficacy Guidelines (2) describe specific requirements for clinical studies.

PRODUCT DEVELOPMENT

Product development commences after lead candidate selection, continues throughout the clinical development stage, and leads to regulatory submissions. Product and manufacturing process understanding is increasingly gained as the clinical development progresses. The clinical batches evaluated in early clinical studies, such as Phase I or pilot bioequivalence studies, are non-final formulations manufactured in small batch sizes using pilot-scale equipment. Following the early clinical studies, the formulation and manufacturing process is further optimized. The batch size increases as the manufacturing equipment moves from pilot-scale to commercial-scale equipment. The batches produced for primary stability and pivotal clinical studies are typically the final formulation and manufactured using the proposed commercial production equipment. Any changes made between the primary stability batches and the intended commercial batches must be justified and supported with product data.

Regulatory agencies require that pharmaceutical companies adopt the principles of Quality by Design (QbD) during the pharmaceutical product development process. The International Council for Harmonization (ICH) Quality Guidelines (3) describe specific requirements for pharmaceutical quality based on Good Manufacturing Practice (GMP) risk management. QbD is a systematic approach that integrates scientific understanding and quality risk management to design, develop, and manufacture pharmaceutical products to ensure consistent product quality and performance, thereby ensuring patient safety and efficacy. Quality should be designed into the product through a robust formulation and manufacturing process. Product and process understanding is gained throughout the development process, and controls are implemented to ensure acceptable product quality at release and throughout the product shelf life. This leads to increased process capability and reduced product variability and defects. The foundations of QbD are described in the ICH Q8 (R2) (4), ICH Q9 (5), and ICH Q10 (6) guidelines and further elaborated by Yu et al. (7) Key concepts of the QbD approach include the quality target product profile (QTPP), critical quality attributes (CQAs), critical material attributes (CMAs), critical process parameters (CPPs), and the control strategy. These terms will be fully described in the sections below.

The QTPP defines the predefined targets for establishing product quality. It summarizes the design requirements of the drug product and includes the intended use, route of administration, dosage form, dose strength(s), container closure system, drug release attributes (e.g., dissolution or aerosol performance), and the critical quality attributes (CQAs) of the drug product. A CQA is a physical, chemical, biological, or microbiological property or characteristic that must be controlled within a specific limit or range to ensure the desired product quality. CQAs are identified by assessing which quality attributes of a drug product will produce severe harm to the patient if the product falls outside the acceptable range. Drug product CQAs generally include identity, assay, content uniformity, impurities, moisture content, and microbial quality. Additional CQAs are specific for different dosage forms, such as dissolution for oral solid products, sterility for parenteral products, and aerosol particle or droplet size distribution for nasal/respiratory products.

During the product design stage, studies are performed to understand the physicochemical properties of the drug substance or active pharmaceutical ingredient (API) and excipients so that the appropriate formulation is identified and optimized. Preformulation studies identify how the API physicochemical properties may impact solubility, manufacturability, and drug release. For example, the salt or polymorphic form may affect the solubility; or particle size may affect manufacturability, dissolution, or aerosol performance. API/excipient compatibility and forced degradation studies identify potential interactions, degradation pathways, and stability issues. The excipient types and grades are selected to aid manufacturability, improve physical or chemical stability, enhance or control drug release and solubility, and improve patient palatability and acceptability. During formulation development, studies are performed where the quantity and grade of the selected excipients are adjusted and optimized to provide a robust formulation that produces the desired drug release and performance characteristics. Critical material attributes (CMAs) are identified for the drug substance, excipients, and in-process materials. A CMA is a physical, chemical, biological, or microbiological property or characteristic of the input material that must be controlled within a specific limit or range to ensure the desired product quality. A material characteristic is deemed critical when a realistic change can have a significant impact on the quality of the output material. Examples of material attributes are provided in Table 8.1. Examples of CMAs may include API polymorphic form, API particle size distribution, excipient particle size distribution, granule particle size distribution, granule moisture content, or powder flow properties. CMAs may affect the manufacturability or product performance at release or on stability.

The pharmaceutical manufacturing process typically consists of a series of unit operations. For example, the manufacturing process for a tablet may include the following unit operations: dry mixing, wet granulation, fluid bed drying, milling,

TABLE 8.1
Examples of Material Attributes

Category	Material Attributes (MAs)	Potential Drug Product CQAs Affected
API	Salt Polymorphic form Particle size and particle size distribution Particle shape Particle true density (porosity) Electrostatic properties Cohesive/adhesive properties Moisture content Potency	Identity Assay Content uniformity Impurities Drug release performance (e.g., dissolution and aerosol performance)
Excipient	Particle size and particle size distribution Particle shape Bulk/tapped density Compressibility Moisture content Impurity content (e.g., metals and residual solvents)	Content uniformity Impurities Drug release performance (e.g., dissolution and aerosol performance)
Granules	API polymorphic form API content and content uniformity Granule size and size distribution Granule shape Granule porosity Granule strength/viscoelasticity Granule impurities Bulk/tapped density Moisture content Residual solvent content	Assay Content uniformity Impurities Drug release performance (e.g., dissolution and disintegration)
Blends	Blend content and content uniformity Particle size and particle size distribution Bulk/tapped density Moisture content Flow properties Segregation potential Electrostatic properties	Assay Content uniformity Impurities Drug release performance (e.g., dissolution and aerosol performance)
Extrudates	API polymorphic form API content and content uniformity Extrudate length, width, and thickness Extrudate mechanical strength Extrudate viscoelasticity Extrudate impurities	Assay Content uniformity Impurities Drug release performance (e.g., dissolution and disintegration)
Tablets	Appearance Weight and weight uniformity API content and content uniformity Tablet length, width, and thickness Tablet hardness Tablet friability Moisture content Disintegration Dissolution Coating thickness Residual solvent content	Appearance Assay Content uniformity Drug release performance (e.g., dissolution and disintegration)

blending, compression, and tablet coating. In a traditional batch manufacturing approach, discrete batches are manufactured by each unit operation resulting in intermediate products. There is generally a time gap between each unit operation. In continuous manufacturing, raw materials are continuously fed into the manufacturing process with the unit operations integrated into a single continuous process stream. The product moves without interruption through each unit operation, eliminating intermediate products and reducing the processing time. Product and process understanding is required to design and control the manufacturing process regardless of whether a traditional batch or continuous manufacturing approach is employed.

During the process design phase, the manufacturing process is evaluated and optimized so that the desired product quality is reproducibly achieved. Formulation and manufacturing process design selection is typically conducted using lab-scale and pilot-scale equipment. The manufacturing equipment type and process should be appropriate for the intended drug product. During manufacturing trials, the process parameters of the manufacturing equipment are adjusted and optimized. Formulation and process optimization is typically conducted along with batch size scale-up on the proposed commercial-scale equipment. Understanding of the manufacturing process is gained, and critical process parameters (CPPs) are identified for each unit operation. A process parameter is deemed critical when its variability impacts the drug product CQA such that the process parameter needs to be monitored or controlled to produce the desired product quality and performance characteristics.

The container closure system for the drug product is evaluated to ensure product quality is maintained throughout the product's shelf-life. The container closure provides protection against moisture, light, and contamination during storage, transportation, and patient use. The material of construction should be considered, and the primary packaging should be compatible with the drug product. Studies may be required to assess the packaging material for drug adsorption, extractables, and leachables. Desiccants may be required to protect against external moisture ingress. Device performance studies are required if the container closure forms part of the dose delivery (e.g., metered dose inhalers or dry powder inhalers).

In practice, formulation studies, manufacturing, and packaging trials are conducted to select the preliminary formulation, manufacturing, and packaging processes. This preliminary formulation and process may be used to manufacture batches for early clinical studies (Phase I, II, or pilot bioequivalence studies). Based on current product data and prior scientific knowledge of similar products, an initial risk assessment is conducted on the preliminary formulation, manufacturing, and packaging process to identify high-risk material attributes and process parameters. Formulation, manufacturing, and packaging optimization studies are conducted to identify CMAs and CPPs and determine any interactions or relationships between the CMAs and CPPs. The final formulation and manufacturing/packaging process is defined, and a control strategy is developed to ensure product quality is consistently met and maintained throughout the product shelf-life. A final risk assessment is conducted on the final formulation, manufacturing, and packaging process to demonstrate that the proposed control strategy adequately controls the high-risk material attributes and process parameters. This final formulation and manufacturing/packaging process is typically used to manufacture batches for primary stability batches and pivotal clinical studies (Phase III or pivotal bioequivalence studies).

The control strategy is based on the product and process understanding gained through pharmaceutical development studies, along with prior knowledge from similar products manufactured with similar manufacturing processes. The control strategy includes the control of material attributes of the drug substance, excipients, in-process or intermediate materials, controls for manufacturing and packaging equipment and process parameters, controls for packaging materials, and in-process and finished product testing at release and on stability.

Depending on the level of product and process understanding gained during development studies, the control strategy may range from Level 3, based on a minimal QbD approach, to Level 1, based on an enhanced QbD approach (Yu et al.). The minimal/Level 3 approach is the traditional approach used in the pharmaceutical industry. The product and process understanding is based mostly on empirical studies, where one factor is adjusted and evaluated. The CMAs and CPPs are highly constrained. The manufacturing process is fixed, and in-process control tests are generally performed off-line. The product specifications and in-process and finished product testing are the primary means of control. In contrast, the enhanced/Level 1 approach utilizes automatic engineering control (e.g., Process Analytical Technology) for real-time monitoring of input material attributes and allows adjustment of process parameters within a design space. Product and process understanding is gained through multivariate design of experiment studies, leading to a comprehensive understanding of the impact of CMAs and CPPs on the drug product CQAs. The product control is shifted upstream and allows for real-time release or reduced finished product testing. Either minimal or enhanced QbD approaches are accepted for regulatory submissions.

A visualized example of the control strategy for an oral tablet product is provided in Figure 8.2. For this example, the drug product CQAs include appearance, identity, assay, content uniformity, impurities, and dissolution. Appearance is controlled by the CMAs of tablet appearances and the CPP of compression force. Identity is controlled by API identity. The assay is controlled through the CMAs of API potency and tablet weight. Content uniformity is controlled by the CMAs of API particle size distribution, blend uniformity, blend particle size distribution, tablet weight uniformity, and the CPPs of blending speed and time. Impurities are controlled by the CMAs of API impurities, antioxidant quantity, granule moisture content, blend moisture content, and the CPPs of seal integrity and desiccant quantity. Dissolution is controlled by the CMAs of API salt, API polymorphic form, API particle size distribution, binder quantity, disintegrant quantity, tablet hardness, and the CPPs of granulation water quantity, kneading time, milling screen size, lubrication speed, lubrication time, compression force, seal integrity, and desiccant quantity. The CMAs of input material may be controlled through raw material specifications and testing. The quantity of input materials is controlled by the master batch records and dispensing records. The CPPs are controlled by the master batch records. CMAs of the intermediate products (e.g., granules and blends) may be controlled through

Input Materials	Granulation	Drying	Milling	Blending	Compression	Packaging	Drug Product CQAs
					Compression force Tablet appearance		**Appearance**
API identity							**Identity**
API potency					Tablet weight		**Assay**
API particle size distribution				Blending speed/time Lubrication speed/time Blend uniformity Blend particle size distribution	Tablet weight uniformity		**Content Uniformity**
API impurities Antioxidant quantity		Inlet temperature Granule moisture content		Blend moisture content		Seal integrity Desiccant quantity	**Impurities**
API salt API polymorphic form API particle size distribution Binder quantity Disintegrant quantity	Water quantity Kneading time		Screen size	Lubrication speed/time	Compression force Tablet hardness	Seal integrity Desiccant quantity	**Dissolution**

Legend:
Critical material attributes (CMAs)
Critical process parameters (CPPs)

FIGURE 8.2 Visualized example of a control strategy for an oral tablet product.

in-process specifications and testing. The product quality may be verified through finished product specifications and testing.

Validation of the manufacturing process is required prior to commercial production. Process validation is defined as the documented evidence that the process, operating within the established parameters, can perform effectively and reproducibly to produce a drug product that meets its predetermined specifications and quality attributes. Process validation confirms that the control strategy is adequate. Process validation can be performed using traditional, continuous verification, or hybrid approaches. The principles of process validation are provided in FDA and EMA guidances on process validation.

The traditional process validation is performed after the final formulation and manufacturing process has been defined. Three consecutive batches are manufactured using the proposed commercial manufacturing equipment at the proposed commercial manufacturing site(s). The manufacturing equipment must be qualified and within the calibration period for the process validation exercise. Although process validation does not need to be completed before regulatory submission, the process validation protocol is included in the dossier and must be completed prior to manufacturing for commercial marketing. The validation protocol should include a short description of the manufacturing process; the stated CPP setpoint and operating ranges; references to finished product specifications and analytical methods; in-process control tests with acceptance criteria; any additional tests with acceptance criteria; and the sampling plan with sampling points, quantity, and frequency. In general, process validation batches have greater levels of sampling and testing with higher scrutiny than typically used for routine commercial production. This is to ensure that acceptable product quality is present throughout the entire batch. A report is issued after the process validation is completed, which summarizes the in-process control, additional tests, and finished product results and conclusions. Any modifications or changes from the process validation protocol should be included with appropriate rationale. If the process validation is conducted prior to regulatory submission, the process validation report is included in the dossier. The process validation protocol and report are also known as the process performance qualification (PPQ) protocol and report.

The continuous verification approach is based on the product and process knowledge gained during pharmaceutical development, supplemented with historical manufacturing data and experience from similar products. Product data from pilot-scale, batch scale-up, and process optimization batches can be used to provide assurance that the manufacturing process performs as expected. The cumulative data from relevant batches produced from the manufacturing process demonstrates that the process operating with the predefined parameters produces material that meets its CQAs and control strategy requirements. A monitoring system is established to provide assurance that the process remains in a state of control during the production of commercial batches.

Specifications are defined in the ICH Q6A guidance (8) as a set of criteria that the drug substance or drug product should conform to be considered acceptable for its intended use. Specifications contain the analytical test, qualitative or quantitative acceptance criteria, and a reference to the analytical method. Specifications are used to define the acceptable range or limits for the CQAs of the drug product and serve as quality standards that must be met to ensure product quality and compliance with regulatory requirements. Specifications are proposed and justified by the manufacturer in the regulatory submission. The justification of specifications contains relevant development data, pharmacopeial standards, and batch data from clinical and primary stability batches at release and from accelerated and long-term stability conditions. The regulatory agency reviews the proposed specifications, supporting data, and justifications. At times, the regulatory agency may request more restrictive acceptance criteria than the manufacturer initially proposed. The specifications approved by the regulatory agency must be complied with by the manufacturer. It is noted that different regulatory agencies may approve differing specifications for the same drug product. Thus, manufacturers must comply with each specific country's approved specifications.

Universal tests for drug substances include description, identification, assay, and impurities (organic, inorganic, and residual solvents). Specific tests for drug substances may include physiochemical properties, such as melting point, particle size, polymorphic form, chiral purity, water content, and microbial limits. Testing may be conducted by solid-state IR, solid-state NMR, Raman spectroscopy, x-ray powder diffraction, and thermal analysis (e.g., DSC and TGA).

Universal tests for drug products include description, identification, assay, degradation products (individual specified products, unidentified products, and total degradation products), and residual solvents. Process impurities from the drug substance synthesis are normally controlled during drug substance testing and are not included in the total impurities of the drug product. If a synthetic impurity is also a degradation product, it is usually included in the total degradation product limit. The specific tests for different dosage forms are described in ICH Q6A guidance and pharmacopeial general monographs. These are summarized in Table 8.2.

Pharmacopeial monographs provide harmonized test procedures. Commonly referred monographs include Uniformity of Dosage Units (USP <905> (17), Ph Eur 2.9.40), (16), Disintegration (USP <701> (18), Ph Eur 2.9.1 (19)), Dissolution (USP <711> (20), Ph Eur 2.9.3 (21)), Aerosols, Nasal Sprays, Metered-Dose Inhalers, and Dry Powder Inhalers (USP <601> (22)), Bacterial endotoxins (pyrogens) (Ph Eur 2.6.14 (23)), Microbial Enumeration Tests (USP <61 (24)), and Tests for Specific Microorganisms (USP <62> (25)).

Product-specific analytical methods are developed and validated to ensure they are fit for the intended purpose. Identification tests are qualitative tests that ensure the identity of the active substance in a sample. Assay tests are quantitative tests measuring the amount of active substance in a sample. Assay tests may be used in conjunction with other tests, such as content uniformity or dissolution. Impurities tests may be a quantitative test or limit test for the impurity in the sample and reflect the purity characteristics of the sample. Analytical methods are validated to verify the following: accuracy, precision (repeatability, intermediate precision), specificity, limit of detection, limit of quantification, linearity, range, robustness, and system suitability. ICH Q2(R1) guideline (26) provides further information on analytical method validation.

TABLE 8.2
Specifications Tests for Various Dosage Forms

Dosage Form	Specification Tests	References
Oral products (E.g., tablets, capsules, and granules)	Uniformity of dosage units, disintegration, dissolution, tablet hardness/friability, tablet breaking force, water content, microbial limits, volatile content, and fineness of dispersion (dispersible tablets)	ICH Q6A, USP (10)
Oral liquids, including powders for reconstitution as oral liquids	Uniformity of dosage units, pH, microbial limits, antimicrobial preservative content, antioxidant preservative content, extractables, alcohol content, dissolution, particle size distribution, suspendability, redispersiblity, rheological properties (viscosity/specific gravity), reconstitution time, and water content	ICH Q6A, USP (9)
Parenteral products	Uniformity of dosage units, pH, sterility, bacterial endotoxins, pyrogens, foreign and particulate matter, particulate contamination (sub-visible particles), container content, water content, antimicrobial preservative content, aluminum content, extractables, osmolarity, particle size distribution, redispersibility, reconstitution time, clarity of solutions, loss on drying and functional testing of delivery systems	ICH Q6A, USP, Ph Eur 2.9.19 (11), Ph Eur 2.6.1 (12)
Transdermal and topical products	Uniformity of dosage units, water content, microbial limits, antimicrobial preservative content, antioxidant content, sterility, pH, particle size, crystal formation, minimal fill volume, and apparent viscosity	USP (13)
Otic products	Microbial enumeration tests, tests for specific microorganisms, antimicrobial content, and antimicrobial effectiveness	USP (14)
Ophthalmic products	Foreign and particulate matter, sterility, particle size distribution, antimicrobial preservative content, pH, minimal fill, osmolality and osmolarity, surface tension, viscosity, zeta potential, and bacterial endotoxins	USP (14)
Metered dose inhalers	Foreign particulate matter, leachables, spray pattern, plume geometry, microbial limit, alcohol content, net fill weight, leak rate, uniformity of delivered dose, and aerodynamic particle size distribution	USP <5> (15)
Dry powder inhalers and nasal powders	Content uniformity (pre-metered), net content (device-metered), foreign particulate matter, microbial limit, residual solvents, leachables, and aerodynamic particle size distribution	USP <5> (15)
Nasal sprays	Preservative content, content uniformity (pre-metered), net content (device-metered), particle size distribution (suspensions), spray pattern, plume geometry, pH, osmolality, viscosity, leachables, foreign particulate matter, sterility, microbial limit, and aerodynamic particle size distribution	USP <5> (15)
Inhalation solutions (for nebulization)	Preservative content, stabilizing excipient content, sterility, particulate matter, pH, osmolality, net content, weight loss, leachables, particle size distribution (suspensions)	USP <5> (15)

REGULATORY SUBMISSIONS

Regulatory dossiers for drug products are submitted to regulatory authorities for marketing authorization. The regulatory dossier should contain comprehensive information and supporting data on the drug product's quality, safety, and efficacy, including the manufacture, specifications, and testing of the drug substance and drug product. Regulatory agencies review dossiers to ensure compliance with regulatory guidelines, standards, and requirements, including those described in ICH guidelines, pharmacopeial standards, and regional regulations. The product data should provide sufficient evidence to demonstrate the drug product's quality, safety, and efficacy. During the regulatory review, the regulatory agency may request further information and supporting data, or request changes to be made within the dossier. This may require additional studies to be conducted.

The ICH common technical document (CTD) is a harmonized format that has been accepted by many regulatory agencies. The CTD consists of five modules. Module 1 contains region-specific information that is not contained in the other modules. Each regulatory agency may have different requirements for Module 1. For example, executed batch records are included in Module 1 for certain regulatory authorities. Modules 2, 3, 4, and 5 are common for all regions. Module 2 contains the summaries of Module 3 (Quality), Module 4 (Non-clinical/Safety), and Module 5 (Clinical). These summaries provide the reviewer with a succinct but comprehensive overview of each module and any integration between quality, non-clinical, and clinical modules. For example, toxicology studies described in the non-clinical module may qualify drug impurities identified in the quality module. Module 3 (Quality) contains the drug substance and product information and supporting data. This will be described in greater detail below. Module 4 (Non-clinical/Safety) contains the evaluation of pharmacology (including adsorption, distribution, metabolism, and elimination), pharmacokinetics, and toxicology (including genotoxicity, carcinogenicity, and reproductive and developmental toxicology) of the pharmaceutical. Module 5 (Clinical) contains the pharmaceutical product's biopharmaceutics, clinical pharmacology, efficacy, and safety. It also contains the product rationale and the risk/benefit analysis.

Module 3 (Quality) contains separate sections for drug substance (sections 3.2.S) and drug product (section 3.2.P). There are multiple sections that contain the drug substance and drug product information, manufacturing process understanding, and controls. Drug substance section 3.2.S.1 contains the general information, chemical structure, and physiochemical properties of the API. Section 3.2.S.2 includes a flow diagram and descriptive narrative of the API synthetic route including starting materials, intermediates, solvents, catalysts, and reagents with expected yields. The process controls, equipment, and operating conditions (e.g., temperature, pressure, pH, and time) are provided for each critical step, along with tests, specifications, and acceptance criteria for isolated intermediates. The raw material specifications are provided for starting materials, solvents, and reagents. Section 3.2.S.3 includes characterization, elucidation of structure with spectral analyses, and discussion on stereochemistry and potential isomerism or polymorphism. Section 3.2.S.4 includes the API specifications, analytical procedures, batch analyses, and justification of API specifications. Section 3.2.S.4 describes the reference standards and materials. Section 3.2.S.6 includes a description, materials of construction, and specifications for the container closure system for the API. Section 3.2.S.7 includes accelerated and long-term stability data for the API and proposed storage conditions and retest date.

Drug product section 3.2.P.1 contains a description of the drug product including route of administration, dosage form, and a formula table listing the API and excipients. Section 3.2.P.2 describes the pharmaceutical development history including the design of the formulation, manufacturing process, and container closure system. The product and process understanding obtained during the drug product development studies is described. This includes API/excipient compatibility; formulation and manufacturing process development studies; selection of excipient type and grades; identification of CMAs of the API, excipients, and intermediate products; selection of the manufacturing process and equipment, identification of CPPs; and the development of the control strategy to ensure the drug product CQAs are met. The manufacturing process and batch size of the batches used for clinical studies are described, highlighting differences in clinical batches. Section 3.2.P.3 includes the formula table for the specified batch quantity. A flow chart describes the critical steps and process controls. The descriptive narrative of the manufacturing process includes the equipment type, manufacturing capacity, environmental controls, and the operating ranges for critical process parameters. The control of the critical steps and intermediates is also described. Section 3.2.P.4 contains the specifications and analytical methods for the excipients. Section 3.2.P.5 contains the specifications, analytical methods, batch analyses, and justification of specifications for the drug product. Section 3.2.P.6 describes the reference standards and materials. Section 3.2.P.7 includes a description, materials of construction, and specifications for the container closure system for the drug product. Section 3.2.P.8 includes accelerated and long-term stability data for the drug product and proposed storage conditions and shelf-life.

COMMERCIAL PRODUCTION

After the regulatory approval, the drug substance and drug product batches must be manufactured in accordance with the registered details described in the marketing authorization for each country. This includes sampling and testing the raw materials, active pharmaceutical ingredients (APIs), packaging materials, intermediates, and bulk and packaged products to ensure compliance with the approved specifications. APIs, raw materials, and packaging materials are received with the vendor certificate of analysis. Incoming testing is performed prior to release from quarantine for use in production. The drug products are manufactured based on the approved manufacturing process with specified formulae, equipment, and process parameters. Each batch is documented in the batch manufacturing and packaging records. In-process, release, and stability testing are conducted against the approved methods and specifications. Retention samples are taken and stored for future testing if required for investigations. A typical flowchart of commercial manufacturing is provided in Figure 8.3.

During commercial production, pharmaceutical companies are expected to monitor process performance and product quality to ensure a state of control is maintained and the desired product quality is reproducibly achieved. This requirement is described in the ICH Q10 guidance. Process capability is a statistical measure that quantifies the ability of a manufacturing process to consistently meet the predefined specifications

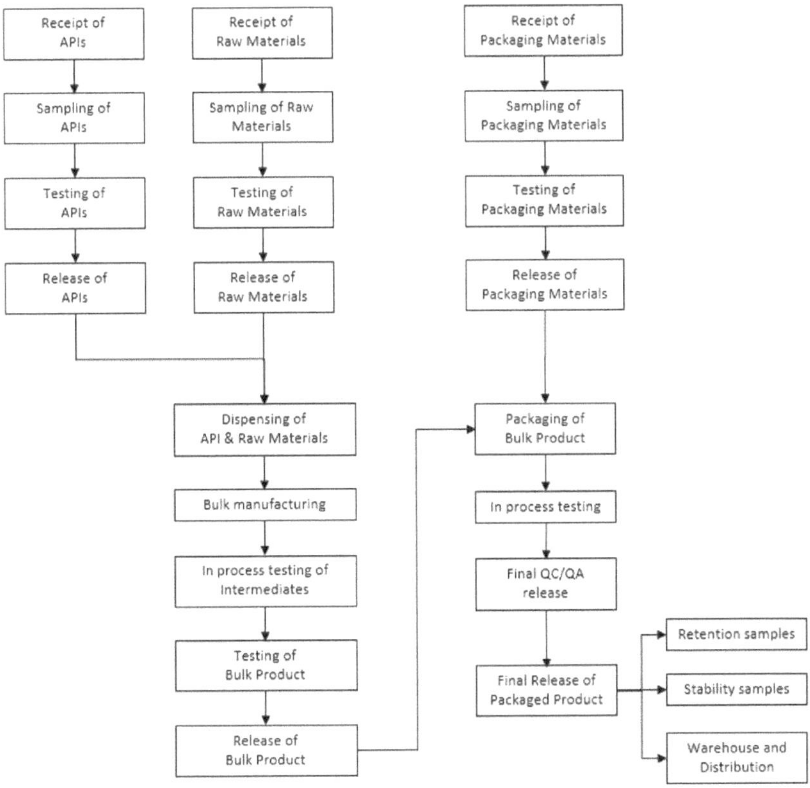

FIGURE 8.3 Typical flowchart of commercial manufacturing.

of the product. It provides an indication of how well the process is performing and is typically expressed by process capability indices, such as Cpk or Ppk values, which compare the spread of the process variability against the width of the specifications. Higher values of Cpk or Ppk indicate better process capability and performance. Cpk values >1.33 are generally acceptable and indicate that the process is under statistical control and performance is behaving as expected. Acceptable Ppk values vary depending on whether the process is centered within the specification limits.

The process capability data are used to distinguish between common cause and special cause variations. Common cause variation is also known as random or systemic variation and is inherent to the process. It represents the normal variation over time and typically follows a normal statistical distribution. In contrast, special cause variation arises from external factors or events that disrupt the normal functioning of the process. Special cause variation is also known as assignable cause variation and leads to sporadic and intermittent changes in the process performance and may result in atypical or out-of-specification results and product defects. Special cause variation is detected when Cpk and Ppk values drop below the acceptable limits. A root cause investigation is performed to identify the source of variation. Corrective and preventative actions are implemented to bring the process back to a state of statistical control.

Any proposed changes should be evaluated against the registered details of the marketing authorization to determine whether regulatory notifications or prior approval submissions are required. There are various reasons for the proposed changes. Some changes are driven by external sources, such as requested changes from suppliers of raw materials, excipients, or packaging materials. Other changes are driven by internal sources, including new suppliers, new equipment, or manufacturing site changes.

Any proposed formulation or manufacturing process changes must be evaluated for potential impact on the product quality. Some changes require regulatory notification after implementation, while others require prior approval before implementation. The specific requirements for post-approval changes vary between regulatory agencies. In general, changes with a greater risk of potential impact on product quality and performance require prior approval before implementation. This general concept is described in ICH Q12 guidance (32). There are various regulatory guidances that outline the category of the intended change, the regulatory requirement for notification or prior approval submissions, and the supporting data requirements. Examples include the FDA Scale-Up and Post-Approval Changes (SUPAC) guidances. A post-implementation assessment should ensure that the proposed change has not had any unintended impact on the drug product performance or quality.

In general, a proposed change with low risk and likely to have negligible or minimal impact on product quality and performance may be notified to the regulatory agency in a periodic report (e.g., Annual Report) after the change has been implemented. Examples of proposed changes with lower risk include equipment change with the same design and operating principle, a batch size increase of up to 10 times using equipment with the same design and operating principles, or a change in mixing time or speeds within the validated range.

A proposed change that may significantly impact product quality and performance may require prior approval from the regulatory agency before implementation. The supporting data required as justification for the proposed change may include product release and stability data comparing drug product batches manufactured with and without the proposed change. In some cases, the regulatory agency may require a bioequivalence study to be conducted. Examples of proposed changes with higher risk include a batch size increase greater than 10 times with the same design and operating principles, equipment change to a different design and operating principles, and change in unit operation (e.g., wet granulation to dry granulation).

CONCLUSION

Pharmaceutical drug products should be designed, developed, and manufactured to ensure consistent product quality and performance, thus, ensuring patient safety and efficacy. Regulatory agencies require that quality-by-design principles are applied during product design and development. The predefined targets for the drug product design are captured in the quality target product profile (QTPP). The product and process understanding gained during product development leads to the identification of critical material attributes (CMAs) of the drug substance, raw materials, and intermediate products; and critical process parameters (CPPs) of the manufacturing process. A control strategy is developed to ensure that the drug product's critical quality attributes (CQAs) are consistently met and maintained throughout the proposed shelf life. A summary of the pharmaceutical development is included in the regulatory dossier, along with details of the formulation, manufacturing process, specifications, analytical methods, and validation of the manufacturing process. Following regulatory approval, the drug product must be manufactured in accordance with the registered details described in the marketing authorization. The process capability should be periodically assessed to ensure that the manufacturing process is operating within statistical control. Special cause variations should be investigated to identify the root cause. Corrective and preventative actions should be implemented to return the process to statistical control. Any proposed changes for the drug product need to be evaluated and may need regulatory approval prior to implementation.

REFERENCES

1. ICH Safety Guidelines: https://www.ich.org/page/safety-guidelines
2. ICH Efficacy Guidelines: https://www.ich.org/page/efficacy-guidelines
3. ICH Quality Guidelines: https://www.ich.org/page/quality-guidelines
4. ICH Q8(R2) guideline: Pharmaceutical Development
5. ICH Q9(R1) guideline: Quality Risk Management
6. ICH Q10 guideline: Pharmaceutical Quality System
7. Yu et al; AAPS Journal 16(4): 771–783. doi: 10.1208/s12248-014-9598-3
8. ICH Q6A guideline: Specifications: Test Procedures and Acceptance Criteria for New Drug Substances and New Drug Products: Chemical Substances
9. USP Oral Drug Products - Product Quality Tests
10. USP Injections and Implanted Drug Products (Parenterals) – Product Quality Tests
11. Ph Eur 2.9.19 Particulate contamination: sub-visible particles
12. Ph Eur 2.6.1 Sterility
13. USP Topical and Transdermal Drug Products - Product Quality Tests
14. USP Mucosal Drug Products - Product Quality Tests
15. USP <5> Inhalation and Nasal Drug Products – General Information and Product Quality Tests
16. Ph Eur 2.9.40 Uniformity of dosage units
17. USP <905> Uniformity of Dosage Units
18. USP <701> Disintegration
19. Ph Eur 2.9.1 Disintegration
20. USP <711> Dissolution
21. Ph Eur 2.9.3 Dissolution
22. USP<601> Aerosols, Nasal Sprays, Metered-Dose Inhalers, and Dry Powder Inhalers
23. Ph Eur 2.6.14 Bacterial endotoxins – pyrogens
24. USP <61> Microbial Enumeration Tests
25. USP <62> Tests for Specific Microorganisms
26. ICH Q2(R1) guideline: Validation of Analytical Procedures
27. FDA Guidance for Industry - Process Validation: General Principles and Practices, January 2011
28. EMA Guideline on process validation for finished products - information and data to be provided in regulatory submissions, November 2016
29. ICH guidance M4S(R2) CTD on Safety
30. ICH guidance M4E(R2) CTD on Efficacy
31. ICH guidance M4Q(R1) CTD on Quality
32. ICH Q12 Lifecycle Management guideline

9 Clinical Applications of Pharmaceutical Particulate Systems

GENERAL BIOPHARMACEUTICS AND PHARMACOKINETICS

The dosage form or drug formulation contains the active pharmaceutical ingredient (API) and excipients designed to provide the specified dose and to facilitate drug delivery and absorption from the selected route of administration. The most common route of administration is oral. Alternative routes include parenteral, respiratory, nasal, buccal, sublingual, transdermal, rectal, vaginal, ocular, and otic. The dosage form facilitates drug transport from the site of administration to the systemic or local site of action. The dosage form is designed for convenience, ease of use, and patient acceptability to improve patient adherence through preparations such as solids, semi-solids, liquids, or aerosols. Most drug products are immediate-release formulations, where the drug is released and absorbed quickly. Faster absorption typically leads to a faster onset of action (1). Some drug products may be designed to provide controlled or sustained drug release over a longer period. All drug formulations should be designed to provide adequate chemical and physical stability throughout the proposed shelf-life, and during storage, transport, and handling.

The drug bioavailability is a measure of the rate and amount of drug that is absorbed and enters the systemic bloodstream. Although the drug needs to reach the target site of action to produce a clinical response, it is generally assumed that the drug present in the systemic bloodstream can reach the site of action. Intravenous drug administration is defined as 100% bioavailability since it directly enters the bloodstream. The absolute drug bioavailability of a drug delivered by other routes of administration is compared against the drug bioavailability administered by IV injection, expressed as a percentage. The relative bioavailability of a drug compares two different formulations or routes of administration.

The pharmacokinetic disposition of the drug is measured to assess the drug concentration in the systemic bloodstream from the initial absorption to its elimination from the body. Key pharmacokinetic parameters include the maximum concentration in the blood (C_{max}), the time to reach the maximum concentration (T_{max}), the total drug exposure over time (area under the curve, AUC), and the time taken to reduce the drug concentration by half (half-life, $t_{½}$). The drug's pharmacokinetic profile is determined by its absorption, distribution, metabolism, and excretion (ADME) properties. Absorption is the passage of a drug from the site of delivery,

DOI: 10.1201/9781003470151-9

dictated by the route of administration, into the systemic circulation. Absorption is a direct determinant of the drug's bioavailability. Distribution is the transport of drug from the bloodstream to the body tissues and organs. Drug distribution depends on the degree of binding to plasma proteins and the extent of tissue penetration. The volume of distribution (V_d) is a theoretical value that indicates how extensively a drug is distributed throughout the body tissues. High V_d values indicate wider distribution throughout the body tissues. Metabolism is the process where drugs are chemically modified through enzymatic biotransformation. Drug metabolites may be inactive, partially active, or fully active with respect to their pharmacological activity. Prodrugs generally are inactive until they are metabolized to their active form. The liver is a primary site of drug metabolism; however, enzymatic metabolism can also occur in other organs and tissues. Excretion is the process of drug elimination from the body. Drug elimination occurs primarily through the urine via kidney filtration or in the feces via biliary excretion. The degree of drug metabolism and elimination affects the drug concentration in systemic circulation over time and determines the drug's terminal half-life and duration of action.

The drug absorption, bioavailability, and clinical effect represent a complex interaction of three main factors: (1) API physicochemical properties, (2) dosage form (formulation) factors, and (3) physiological conditions. The dosage form and physiological factors are dependent on the selected route of administration discussed in later sections of this chapter.

There are multiple steps involved in the drug absorption process, including the transport of the dosage form from the site of administration to the site of absorption; drug release from the dosage form; dissolution in the surrounding physiological fluids; diffusion to the site of absorption; permeation across the membranes; and transport to the bloodstream, lymphatic system, or local tissues. Although drug absorption is complex and involves multiple processes, it may be simplified to the fundamental processes of drug dissolution and drug permeability.

The rate of drug dissolution can be described by the Noyes–Whitney equation (described in Chapter 7).

The API physicochemical properties affecting dissolution include the intrinsic solubility, molecular weight, chemical functional groups, lipophilicity, salt form, degree of crystallinity, polymorphic form, and the degree of ionization (which is influenced by the log of the acid dissociation constant (pKa) and pH of the surrounding biological fluid). The formulation properties may also affect drug dissolution, including the particle size, degree of particle aggregation, effect of excipients and manufacturing process, and the release properties. Physiological factors affecting dissolution include pH, temperature, surface tension, volume, and composition of biological fluid. Specific formulation factors and physiological conditions associated with each route of administration will be described in more detail in later sections.

Drug permeability across membranes occurs by passive diffusion via paracellular or transcellular pathways, carrier-mediated diffusion, carrier-mediated active transport, and endocytosis mechanisms. Passive diffusion is the most common mechanism for drug penetration across biological membranes, especially for small molecules. Carrier-mediated diffusion and active transport mechanisms recruit specific transmembrane carrier proteins or channels and are specific/selective and saturable processes.

The rate of drug diffusion across a membrane can be described by Fick's first law of diffusion (described in Chapter 7). Since biological membranes consist of lipid bilayers, non-ionized and lipophilic drugs can penetrate membranes by passive diffusion. API physicochemical factors affecting drug permeability include the molecular size, lipid solubility (typically characterized by the oil/water partition coefficient, log P), degree of ionization which is influenced by the log of the acid dissociation constant (pKa), and pH of the surrounding biological fluid. Greater drug permeability is observed with smaller MW, lower logP values, and non-ionized drugs. Weakly acidic drugs have higher permeability when the pH is less than the pK_a, whereas weakly basic drugs have high permeability when the pH is greater than the pK_a (1).

Generally, the drug needs to be dissolved to penetrate the biological membranes. In some cases, solid particles may be transported by endocytosis, where the cell membrane engulfs the drug particle by forming a vesicle. This occurs by pinocytosis and phagocytosis mechanisms.

ORAL DELIVERY

The oral route of delivery to the gastrointestinal tract is the most common route of administration and may be used for both systemic and local drug delivery. This route is convenient, easily accessed, and non-invasive. However, oral delivery may encounter acid/base chemical degradation, enzymatic metabolism, and hepatic first-pass effect leading to reduced bioavailability. Oral delivery provides a delayed onset of action, compared with other routes of administration, including intravenous, sublingual, or respiratory. Some drugs cause gastric irritation and may lead to nausea or gastrointestinal bleeding. Oral administration may not be possible in certain circumstances, for example, where patients are unconscious, vomiting, or experiencing seizures.

BIOLOGY

The primary functions of the gastrointestinal tract are digestion of food, absorption of nutrients, re-absorption of water, and elimination of waste products. The gastrointestinal tract (Figure 9.1) consists of the mouth, esophagus, stomach, small intestine (duodenum, jejunum, and ileum), large intestine (cecum, colon, and rectum), and anus. The liver, gallbladder, and pancreas are accessory organs that aid the digestive and absorptive functions. The oral cavity can be used for drug delivery, such as sublingual and buccal administration (discussed later in this chapter). Orally administered drug products are transported to the stomach. The primary function of the stomach is mechanical and chemical digestion. The gastric fluid consists of dietary food and liquids, saliva, gastric secretions (hydrogen ions, lipase, and pepsin), and refluxed liquid from the duodenum. The pH of the stomach ranges from 1–2 in the fasted state to 3–7 in the fed state. Drug absorption in the stomach is limited due to the relatively small surface area. The stomach contents are released into the small

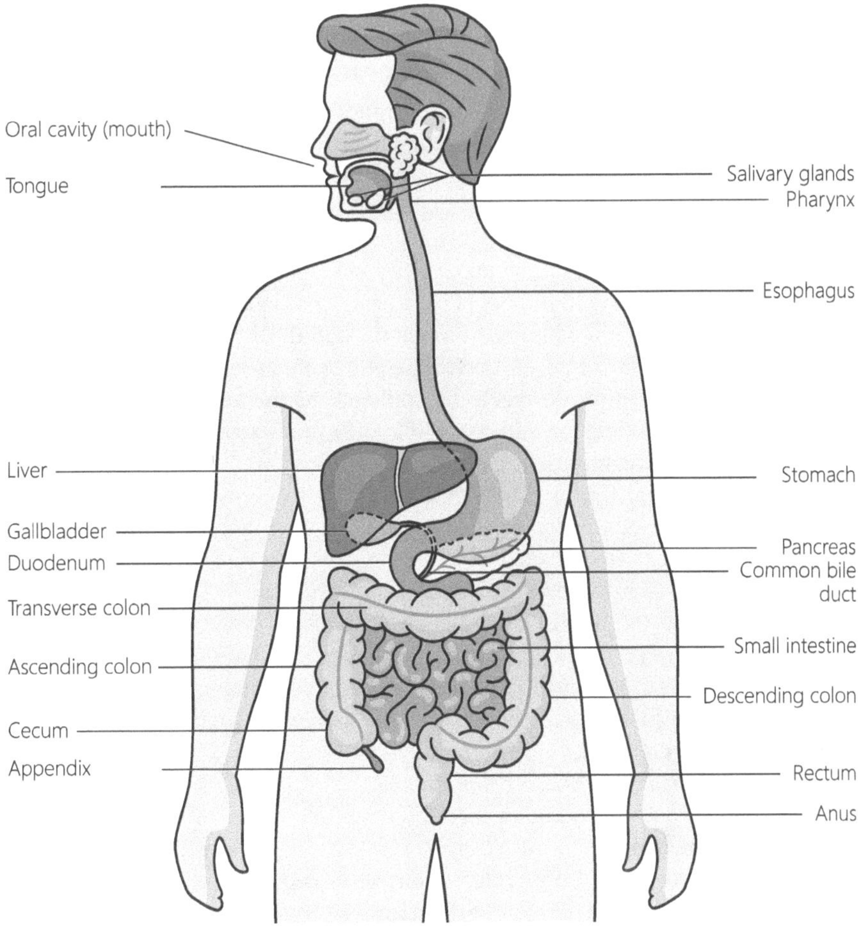

FIGURE 9.1 The gastrointestinal tract (https://www.shutterstock.com/image-vector/gastrointestinal-tract-structure-diagram-hand-drawn-2428221265).

intestine in a controlled manner. The small intestine secretes bicarbonate ions to neutralize the acidic gastric fluids and receives secretions from the pancreas (proteases and amylases) and the liver (bile, phospholipids, cholesterol). The pH of the duodenum is typically 6.5 in the fasted state. The pH increases to 7–8 in the jejunum and ileum. The small intestine is the primary site of drug absorption due to its large surface area (200 m^2), due to the folds of Kerckring, villi, and microvilli. The large intestine is mainly responsible for the absorption of water and electrolytes, and the storage and elimination of fecal material. The pH level in the colon is typically 6–8, but may vary depending on food intake, microbial metabolism, and disease state. Although the enzymatic activity of the large intestine is relatively low, the enzymes from the microflora may cause drug metabolism (1–3). The rectal cavity may be used for drug delivery (discussed later in this chapter).

The mucus layer presents a barrier to drug absorption. The mucus is secreted from goblet cells and is a complex hydrogel composed of mucin, carbohydrates, lipids, salts, antibodies, bacteria, and cellular debris. Mucin is a large molecular weight glycoprotein that provides the hydrogel properties of mucus. The mucus layer functions to lubricate, hydrate and protect the gastrointestinal epithelium against mechanical stress, foreign particles, and pathogens. There is a firmly adherent and a loosely adherent layer, which provides a slippage plane that allows undigested food to be transported through the gastrointestinal tract. The mucus layer thickness varies along the gastrointestinal tract: stomach (189–274 μm), small intestines (120–480 μm), and large intestines (830 μm). A pH gradient exists in the stomach mucus layer, ranging the acidic pH on the luminal surface to neutral pH at the epithelial surface. The mucus layer maintains an unstirred layer throughout the gastrointestinal tract that requires drug diffusion to reach the epithelial membrane prior to absorption. The mucin molecules entangle and cross-link to form a size-exclusion network that entraps pathogens and large particles (4, 5).

The gastrointestinal tract is highly vascularized, receiving 28% of the cardiac output providing a good opportunity for drug absorption into the systemic circulation. "Sink" conditions are present due to the large blood volume and rapid clearance of the drug following absorption into the systemic circulation. This high concentration gradient favors the unidirectional transfer of drugs from the gastrointestinal tract to the blood. The blood flow from the gastrointestinal tract drains into the hepatic portal vein and passes through the liver before entering the systemic circulation (known as the first-pass effect). Drugs susceptible to hepatic metabolism have reduced bioavailability (1, 2, 6–8).

The gastrointestinal tract also has a rich supply of lymphatic vessels. Since the blood flow rate is 500 times higher than the lymph flow rate, most drugs absorbed across the intestinal membrane are transported into the portal blood system. However, lipophilic drugs (log $P > 5$) are preferentially absorbed into the lymphatic system. Drugs with higher molecular weight may selectively absorb into the lymphatic system because lymphatic capillaries are more permeable than blood capillaries. Drugs transported to the lymphatic system bypass hepatic first-pass metabolism since the mesenteric lymph empties directly into the systemic circulation. Lymphatic drug absorption is enhanced by association with colloidal lipoproteins that assemble from digested fatty acids and monoglycerides (9).

Biopharmaceutics

The rate and extent of drug absorption from the GI tract depends on the API physicochemical properties, formulation factors, and physiological conditions. In general, drugs must be dissolved to allow drug absorption, and most drugs are absorbed across the intestinal membrane. There are some cases where drugs may be absorbed in a particulate state, which will be discussed later in this chapter. The drug absorption is affected by the drug release from their oral dosage forms, the dissolution rate in the gastrointestinal fluids, and the residence time in the small intestines.

The API physicochemical properties and formulation variables affecting drug dissolution and permeability impact the oral drug absorption and bioavailability. The API intrinsic solubility is influenced by the molecular weight, chemical functional groups (hydrophilic, hydrophobic, and ionizable groups), degree of crystallinity, and polymorphic form. Lower solubility is observed with larger molecules (higher MW) and increased lipophilicity (higher logP values). Amorphous drugs generally exhibit greater solubility and faster dissolution rates. Some polymorphic forms have higher solubility than others, depending on the molecular packing arrangement and crystal lattice strength. Molecules with higher melting points generally have lower intrinsic solubility. The degree of ionization is determined by the pKa and the pH of the surrounding fluids and affects the aqueous dissolution rate. Ionized drugs exhibit faster dissolution rates. The rate of drug dissolution is affected by the API particle size, shape, surface area, and degree of drug aggregation. Based on the Noyes–Whitney equation (described in Chapter 7), the dissolution rate increases with a higher surface area. Smaller particles generally have a higher relative surface area, which increases the dissolution rate. However, smaller particles may aggregate which reduces the effective surface area. Differently shaped particles exhibit higher surface area (e.g., needles).

Formulation excipients may increase the rate of drug dissolution by reducing the surface tension of the dissolution media (e.g., surfactants and wetting agents). Some excipients may increase drug permeability across intestinal membranes (e.g. fatty acids, surfactants, bile salts, chitosans, and cyclodextrins).

The biopharmaceutical classification system (BCS) is a scientific framework to classify APIs based on their aqueous solubility and intestinal permeability. This concept was originally published by Amidon in 1995 to simplify FDA standards of bioequivalence for orally administered drug (10, 11). This approach has become widely accepted in academics, industry, and regulatory fields. The BCS categorizes drugs into four classes: Class I—highly soluble and highly permeable, Class II—poorly soluble but highly permeable, Class III—highly soluble but poorly permeable, and Class IV—poorly soluble and poorly permeable. A highly soluble drug is defined as a drug where the highest dose strength is soluble in 250 mL of aqueous media in the pH range of 1.2–7.4 (which represents the pH range experienced in the gastrointestinal tract). A highly permeable drug was originally defined as an absolute bioavailability of greater than 90% but has been revised to an absolute bioavailability of greater than 85%. The drug solubility is determined by in vitro dissolution testing using the USP basket method at 100 rpm or the paddle method at 50 rpm in pH 1.2, 4.5, and 6.8 buffer solutions, representing the stomach, duodenum, and upper small intestine. The drug permeability may be determined from human pharmacokinetic studies or validated in vitro methods (e.g., Caco-2 cell permeability assay). An immediate release oral dosage form containing BCS Class I and III APIs may be eligible for a BCS-based biowaiver, which allows in vitro dissolution testing to be conducted instead of in vivo bioequivalence studies (ICH M9 guidance) (11).

The BSC framework was expanded to include BCS Class II sub-classes based on the drug pKa values. The BSC Class IIa sub-class includes weakly acidic drugs with pKa < 5. These APIs have a high degree of ionization in the basic conditions of the small intestine, leading to high solubility and relatively fast absorption across the intestinal membrane. The BSC class IIb sub-class includes weakly basic drugs with pKa > 6. These APIs have a high degree of ionization and high solubility in the acidic conditions of the stomach but may precipitate in the basic conditions of the small intestine. This leads to complex absorption across the intestinal membrane, which is dependent on the formulation and GI physiological factors. The BCS class IIc sub-class includes neutral drugs with pKa < 0 or > 8. The solubility of these non-ionized APIs may be enhanced by bile acids and lipids present in the gastrointestinal tract (12).

The physiological factors affecting drug absorption include the volume, pH, and composition of gastrointestinal fluids; the presence of food; the rate of gastric emptying and intestinal transit; and drug degradation and metabolism.

The volume of liquid across the gastrointestinal tract varies, which in turn affects the rate and extent of drug dissolution. The volume of liquid is affected by the amount of liquid ingested, gastric and pancreatic secretions, gastric emptying rate, intestinal transit time, and uptake or efflux of liquids by the gastrointestinal membrane. In the fasted state, the liquid volume is ~18–54 mL in the stomach, ~30–420 mL in the small intestine, and ~ 13 mL in the large intestine. The pH of gastrointestinal fluids ranges from highly acidic in the stomach to alkaline in the small and large intestines. This affects the degree of ionization of weakly acidic or weakly basic drugs. Ionized drugs exhibit higher solubility but may have lower membrane permeability, whereas non-ionized drugs may exhibit lower solubility but may have higher membrane permeability. Some components of the gastrointestinal fluids may enhance drug solubility and absorption. For example, bile salts increase the solubility and absorption of lipophilic drugs (1, 6, 7, 13).

Gastric emptying determines the drug residence time in the stomach and entry into the small intestine. In the fasted state, gastric emptying is determined by the inter-digestive migrating motor complex (MMC), commonly known as the "housekeeper wave". The MMC is a cyclic motor activity that occurs every 90–120 minutes under fasting conditions. It is characterized by three phases: Phase I is a quiescent period, Phase II has intermittent and irregular contractions, and Phase III has intense contractions where solid materials are released into the small intestine. In the fed state, the MMC is replaced by regular, tonic contractions which mixes food with gastric secretions. Gastric emptying in the fed state is controlled by mechanical, hormonal, and neural mechanisms; and influenced by many factors, such as meal volume, liquid viscosity, osmotic pressure, body position (e.g. lying on the left or right side), emotional state (e.g., depression and stress), and disease state. Liquids and small particles (<7 mm) pass the pyloric sphincter into the duodenum; however, large particles are retained in the stomach until the MMC resumes. Thus, large non-disintegrating dosage forms (e.g., enteric-coated tablets) experience delayed gastric emptying (1, 6, 13, 14).

The intestinal transit and residence time significantly affect drug absorption. Slower transit increases the intestinal residence time and provides a greater opportunity for drug dissolution and absorption. Smooth muscle contractions produce peristaltic waves which propel contents along the gastrointestinal tract and agitate the unstirred mucus layer. Contractions in different segments of the small intestines provide a mixing motion, which may increase drug dissolution due to agitation and may increase drug absorption due to greater contact with the intestinal epithelium. Factors affecting intestinal transit include the presence of food, presence of drugs or excipients (e.g. mannitol reduces gastrointestinal transit time), diet (vegetarian or non-vegetarian), and pregnancy. The measured flow rate in the jejunum increases from 0.73 mL/min in the fasted state to 3.0 mL/min in the fed state. The intestinal transit time is generally 3-4 hours (6).

The presence of food increases peristaltic activity, motility, and secretions in the intestines. The gastrointestinal fluid viscosity is increased when food is present, which reduces the rate of drug dissolution and diffusion to the intestinal membrane. Intestinal secretions are released in response to the presence of food which may lead to greater drug degradation and metabolism (1). The higher buffer capacity and osmolarity of gastric and intestinal fluids observed in the fed state may reduce the rate of drug dissolution. Intestinal secretions produced in the fed state can lower the surface tension and improve wetting and drug dissolution (6). Drugs may become adsorb onto certain foods (e.g, pectins and fibers) or form complexes (e.g., metal ions) (14). Drugs that exhibit reduced absorption in the presence of food should be taken on an empty stomach. Drugs that cause gastrointestinal irritation should be taken with food (15).

A longer residence time is observed in the colon. It is highly variable and may range from 6 to 70 hours. The transit time in the large intestine is influenced by diet, fasted/fed state, mobility, stress, and colon diseases (e.g., ulcerative colitis and irritable bowel syndrome). A high variety of microbiota (>400 aerobic and anaerobic species) is present in the colon which contributes to the enzymatic and chemical drug degradation. Limited drug absorption occurs from the colon due to the low surface area, low fluid volume, highly viscous media, variety of microbiota, mucus layer, and high individual variability (1, 7). While rectally administered dosage forms can reach the rectum and descending colon, oral administration is required for drug delivery to the ascending and transverse colon. Some orally administered drug formulations target local delivery to the colon, including prodrugs activated by colonic microflora, pH-sensitive polymers, time-controlled release systems, bacteria-degradable polymers, and osmotic-controlled systems (7, 16, 17).

Drug degradation and metabolism may occur at various sites along the gastrointestinal tract resulting in reduced drug absorption. Some drugs may degrade in the highly acidic conditions in the stomach or alkaline conditions in the small intestines. Peptides and proteins may be inactivated by enzymes, such as pepsin in the stomach and trypsin from pancreatic secretions. Enzymatic and acid/base-mediated degradation may occur within the gastrointestinal fluids, by mucosal cells lining the gut walls, and by microorganisms within the colon (1).

Drug absorption mostly occurs in the small intestine, especially in the duodenum and jejunum, due to the large surface area and extensive capillary network. Absorption pathways include passive diffusion via the paracellular or transcellular pathway, active influx transporters, or vesicular endocytosis. The intestinal membrane also contains active efflux transporters (e.g., P-glycoprotein) which transport drugs into the intestinal tract resulting in drug clearance and elimination (6, 18).

Drug absorption by particulate uptake may occur through gut-associated lymphatic tissue (GALT) (19, 20). Particulate uptake occurs by phagocytosis (e.g., M cells, macrophages, monocytes, neutrophils, and dendritic cells) or pinocytosis. There are various types of pinocytosis, including micropinocytosis (particles > 1 micron), clathrin-mediated endocytosis (particles ~120 nm), caveolin-mediated endocytosis (particles ~60 nm), and clathrin- and caveolin-independent endocytosis (particles ~90 nm). Small particles <50–100 nm are taken up by enterocytes and are transported to the systemic bloodstream. Larger particles >100 nm are taken up by M cells and are transported to the lymphatic system, which bypasses hepatic first-pass metabolism (21, 22). The rate of particle uptake is affected by the physicochemical properties of the drug, including hydrophobicity and particle size (19). Smaller particles exhibit a higher rate and extent of particle uptake (23, 24).

Preparations

Oral dosage forms may be liquid or solid preparations. Liquid preparations include aqueous solutions, suspensions, and oil-in-water emulsions. Solid preparations include tablets and capsules. In general, drug bioavailability follows (in descending order): solutions, suspensions, oil-in-water emulsions, capsules, tablets, and modified-release tablets/capsules. Aqueous solutions may be absorbed upon reaching the duodenum, whereas suspensions must dissolve prior to absorption; tablets and capsules must disintegrate prior to drug dissolution and absorption (15).

Techniques to improve oral bioavailability of poorly water-soluble drugs include API physiochemical modifications or formulation technologies, including particle size reduction and novel technologies.

API physicochemical modifications have been used to improve drug solubility and oral bioavailability of poorly water-soluble drugs. These approaches include changes in the salt form, degree of crystallinity, degree of solvation, polymorphic form, and pro-drug conversion by esterification to reduce lipophilicity and co-crystallization to increase solubility (13, 25).

Formulation technologies are used to improve the drug solubility, dissolution, and oral bioavailability of poorly aqueous soluble drugs. Particle size reduction has been historically used to improve the solubility and dissolution rate of drugs. Particle size reduction results in greater surface area and a faster rate of dissolution (26–28). Increased dissolution and bioavailability with particle size reduction have been observed for many poorly soluble drugs. However, aggregation of small particles may increase the effective surface area, resulting in unchanged or reduced dissolution rate and bioavailability (29, 30). Reduced aggregation of drug particles and increased dissolution rates may be obtained by using solid dispersions (31, 32) and interactive mixtures (33, 34).

Newer technologies to improve drug bioavailability include: self-emulsifying drug delivery systems (SEDDS), complexation with cyclodextrins, lyophilized liposomes, solid lipid nanoparticles, and amorphous solid dispersions.

SEDDS are lipid-based formulations designed to improve the oral bioavailability of poorly water-soluble drugs. SEDDS may be liquid or solid formulations and generally contain lipophilic drugs, lipids (e.g., fatty acids, vegetable oils, or derivatives), surfactants (e.g., Tween 80), and co-solvents (e.g., ethanol and glycerol) encapsulated in an enteric-coated capsule. SEDDS formulations release in the small intestine and spontaneously form a microemulsion upon contact with the intestinal fluids, containing bile salts and digested lipid components (e.g., cholesterol, monoglycerides, and fatty acids). The nanosized droplets allow drug absorption across the intestinal membrane and uptake into the lymphatic system. SEDDS enhance drug permeability by lipid fluidization of intestinal membranes and opening of tight junctions. SEDDS may also enhance bioavailability by inhibition of P-gp efflux. The SEDDS formulations may be manufactured by adsorption onto solid carriers, melt granulation, melt extrusion, spray-cooling, or spray-drying (4, 35–37).

Cyclodextrin formulations have been used for oral and parenteral formulations. Cyclodextrins are cyclic oligosaccharides derived from starch and consist of a hydrophilic outer surface and a lipophilic central cavity. They may assemble into various structures including micelles, nanoparticles, nanorods, and other complexes. Hydrophilic drugs may be incorporated into the aqueous core of the cyclodextrin assembly or the hydrophilic exterior, whereas hydrophobic drugs may be incorporated in the hydrophobic cavity. Cyclodextrin complexes protect API from chemical degradation, enhance membrane permeability, and may facilitate sustained drug release (37, 38).

Liposomes are spherical vesicles containing a phospholipid bilayer that encases an aqueous core. Liposomes can encapsulate both hydrophilic and lipophilic drugs. Hydrophilic drugs are encapsulated in the aqueous core, whereas lipophilic drugs are integrated within the lipid bilayer. Although liposomal formulations are generally liquid preparations, they may be spray-dried or freeze-dried with stabilizers (e.g., trehalose) to provide liposomal powders for tablet or capsule formulations (37).

Amorphous solid dispersions (ASDs) are a supercooled glass material that remains in a state of metastable thermodynamic equilibrium, stabilized with polymers (e.g., hypromellose, PVP, copovidone, and Eudragit). The drug forms a molecular dispersion with the polymeric substrate resulting in a lowered glass transition temperature (T_g). Additional excipients may be added as processing aids and additional stabilizers, including plasticizers, surfactants, and antioxidants. ASDs are most commonly manufactured by spray-drying or hot melt extrusion. The ASD intermediate may be formulated into powder for sachets, tablets, or capsules (25).

Nanoparticulate technologies may be used to target particulate uptake via the GALT pathway. Examples include polymeric nanocapsules or nanospheres, and solid lipid nanoparticles. Polymeric nanocapsules are vesicular systems where the drug is enclosed in a cavity surrounded by the polymer membrane. Polymeric nanospheres contain drug dispersed in the polymer matrix. The polymer may be natural (e.g., dextran, alginate, agar, and chitosan) or synthetic (e.g., PLA, PGA, PLGA, and PCL). These polymeric nanoparticles may include surface modifications to target specific cells or receptors; or are mucoadhesive for prolonged retention time. They have been investigated for oral delivery of poorly soluble drugs, peptides, proteins, and vaccines. Nanoparticles are generally <200 nm to allow diffusion through mucus to the intestinal membrane. Particle uptake occurs via phagocytosis or pinocytosis (21, 22).

Solid lipid nanoparticles (SLNs) are colloidal drug carrier systems comprised of a solid lipid core, stabilized by surfactants. The drug is incorporated within the lipid matrix or adsorbed onto the particle surface. SLNs are solid at both room and body temperatures. SLNs are used to improve the bioavailability of poorly soluble drugs and may be used for oral or parenteral delivery. Following oral administration, they are absorbed into the lymphatic system and bypass hepatic first-pass metabolism. The SLN intermediates can be compressed into tablets or capsules for oral delivery (36, 37, 39). SLNs appear to be taken up by pinocytosis mechanisms, depending on their particle size (22).

PARENTERAL DELIVERY

Parenteral drug delivery may provide systemic action or local action, depending on the injection site and formulation. The parenteral route of administration allows direct drug delivery into a body compartment, such as blood vessels, organs, tissues, or lesions. The most common routes of administration for parenteral products are intravenous (IV), intramuscular (IM), subcutaneous (SC), and intradermal (Figure 9.2). Other parenteral routes include intra-arterial (artery), intrathecal (cerebrospinal fluid), intra-articular (joint), intra-ocular, intra-cardiac, intra-peritoneal, and intra-tumoral. Parenteral injections may be immediate- or extended-release dosage forms.

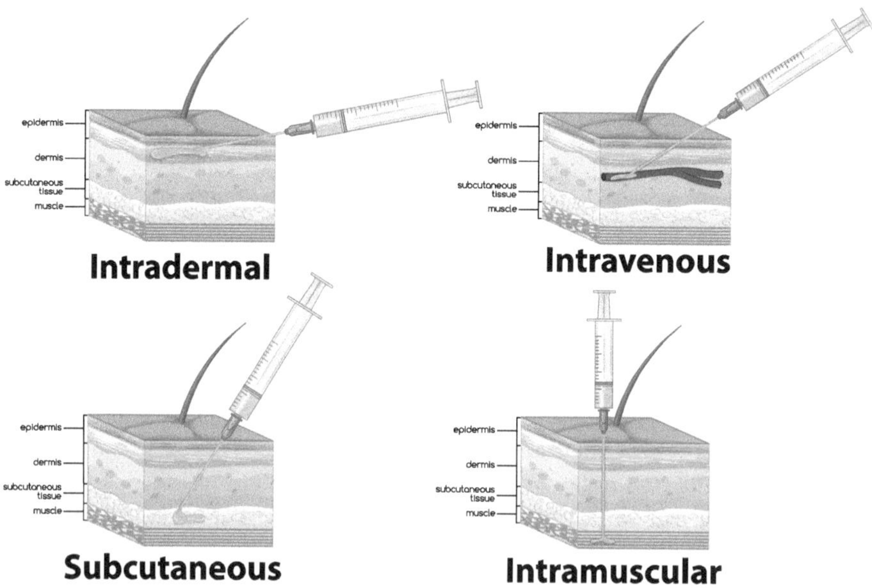

FIGURE 9.2 Common routes of parenteral delivery (https://www.shutterstock.com/image-vector/set-type-injections-illustration-1125078341).

Biology

Intravenous (IV) administration provides an immediate onset of action; whereas other parenteral routes provide slower onset and/or prolonged duration. IV preparations are typically solution-based systems. Suspensions and other particulate systems may be administered via IM or SC injections (15).

Intramuscular (IM) injections are delivered into the large, well-vascularized muscles, including the deltoid (upper arm), vastus lateralis (outer thigh), and dorsogluteal muscle (buttocks). The injection volume ranges from 1 to 5 mL, depending on the muscle size. Smaller volumes are delivered to the deltoid, whereas larger volumes can be accommodated by the vastus lateralis and dorsogluteal muscles. The deltoid muscle has slower absorption than the gluteal and thigh muscles (40).

Subcutaneous (SC) injections are delivered to the layer of adipose and connective tissue just beneath the skin. This tissue is less vascularized compared with muscle, which results in slower and more sustained drug absorption. The sites for SC injection include the upper arm, thigh, abdomen, or upper buttocks. The injection volume for SC injections are smaller volumes (<2 mL), than those used for IM injection. The needles used for SC injections are shorter and finer, compared to those used for IM injections.

Biopharmaceutics

Drug absorption following IM or SC injections requires drug dissolution followed by passive diffusion through the interstitial tissue to the nearby capillaries before reaching the systemic circulation. This results in an initial lag period between the time of injection and the time of entry into the general circulation (41). For solution-based formulations, drug precipitation may occur at the injection site due to dissipation of the aqueous vehicle or change in pH (42). For suspension-based formulations, the liquid vehicle is absorbed into the surrounding tissues resulting in drug particle aggregation and the formation of a depot with reduced surface area for drug dissolution and absorption. The depot formed at the injection site acts as a drug reservoir providing prolonged drug release and duration of action (43).

The IM and SC injections may be immediate-release or long-acting formulations. The duration of drug activity depends upon the physicochemical and pharmacokinetic properties of the drug, degree of plasma protein binding, extent of distribution throughout the body, and rate of elimination by metabolism and/or excretion (15). Long-acting formulations may exhibit flip-flop kinetics, where the terminal elimination rate is controlled by the absorption rate, rather than the true elimination rate. This results in an apparent elimination rate that is equal to the absorption rate (40, 44).

The drug release and absorption rate from aqueous suspensions are mostly controlled by the dissolution rate, largely dependent on the intrinsic drug solubility in the physiological fluid, particle size, and surface area. Lipophilic drugs have slower dissolution and absorption rates, resulting in longer half-life (44, 45). The rate-limiting step for drug absorption of suspension-based formulations is generally the dissolution of the solid drug particles in the biological fluids at the injection site (46). Suspensions with smaller particle sizes have faster dissolution and subsequent absorption rates due to increased surface area (47). Surfactants may reduce aggregation and increase dissolution and absorption rates. Suspension formulations with higher drug concentration may lead to aggregation and slower dissolution and absorption rates. Suspensions with larger particles exhibit slower dissolution and absorption; resulting in a prolonged duration of action (43, 48). Faster in vitro drug release was observed with depot medroxyprogesterone acetate (DMPA) injectable suspension with smaller particles. Higher C_{max} was observed with an entecavir suspension using smaller particles in the rat PK study. A shorter half-life was observed with a cabotegravir injectable suspension with smaller particles in a human PK study (44, 49). Stability studies should be conducted on suspension-based formulations to observe for sedimentation, resuspendability, and particle size growth due to Ostwald ripening, secondary nucleation, or particle agglomeration (50).

Drug absorption may also occur via macrophage phagocytosis or into the lymphatic system (42, 44, 45). The absorption pathway of injected drugs depends on the molecular weight, lipophilicity, and particle size. Molecules smaller than 5 kDa favor absorption into capillaries, whereas larger particles greater than 16 kDa are preferentially drained into the lymphatic system. Lipophilic drugs or formulations, such as micelles, liposomes or emulsions, are taken up by the lymphatic system (42). Macrophages can phagocytose particles smaller than 10 μm, but particles smaller than 200 nm are less efficiently internalized and may escape cellular recognition and uptake. Particles with hydrophobic surfaces are more susceptible to phagocytosis. Drugs engulfed by macrophages travel to the lymphatic system and may accumulate in the lymph nodes, serving as a secondary site for drug release and absorption (44).

Drug absorption depends on the physicochemical properties of the drug (e.g., molecular weight, pKa, solubility, lipophilicity, log P, crystallinity, and polymorphic form), formulation characteristics (e.g., drug concentration, particle size, vehicle composition, pH, viscosity, and osmolarity), injection administration (e.g., injection site, depth, volume, technique, depot shape, needle gauge, and length), and physiological conditions (e.g., blood supply, body movement, tissue transport, lymphatic transport, host response, and gender differences) (15, 40–42, 44–46, 51, 52).

The host response is a complex series of events that may affect the drug absorption of injected suspensions. An injection site reaction is triggered because the IM or SC injection induces tissue injury, and the solid particles are recognized as foreign bodies. The initial transient acute inflammation response leads to edema, which may result in faster drug dissolution during the early inflammatory response. As the host response progresses, the recruitment of inflammatory cells and macrophages may lead to fibrous encapsulation of the drug depot, which reduces drug dissolution and absorption. Sprouting of new capillaries (angiogenesis) into the depot may occur, which promotes drug release through an increased exposed surface. Macrophages may infiltrate the depot and phagocytose drug particles, depending on their particle size. The macrophages with engulfed drug particles subsequently enter the lymphatic system and may accumulate in the lymph nodes, where they serve as a secondary site for drug release and absorption (44). Nanoparticle suspensions may have an extended duration of activity due to macrophage phagocytosis. For example, phagocytosis of cabotegravir nanoparticles from IM injection leads to extended local retention and a longer duration of action (45).

Other factors associated with the drug administration and physiological conditions may affect drug absorption, such as needle length, injection force, injection site, physical activity, body mass index, and gender differences. The injection site, speed, and pressure may affect the level of micro-tearing in the muscle or subcutaneous tissue, which affects the shape and porosity of the drug depot. Drug depots with greater lateral spreading have the higher exposed surface area and faster drug dissolution and absorption. Formulations with lower viscosity show greater lateral spreading (44). MRI imaging may be useful to view the depot size and shape (50).

The needle length determines the depth of the injection. Injection into adipose tissue produces slower drug absorption than muscle tissue due to lower blood flow and water content. Longer needle length may be needed for people with higher body mass index (BMI) to ensure IM injections are delivered to the correct locations. Women have more adipose tissue and less muscle compared with men, which may lead to gender differences in drug absorption from IM injections. Men tend to have higher C_{max} and AUC, and shorter half-lives compared to women. Similarly, muscle mass loss is observed with advancing age, which may affect drug absorption (44).

Physical activity may affect drug absorption from IM depot sites. Increased blood flow to skeletal muscles may increase drug absorption. The pH of the interstitial fluid can drop from 7.4 to 6.8 due to lactic acid production which may accumulate for several hours following exercise. This change in pH may affect the dissolution of drugs with pKa in the physiological range. For example, risperidone has a pKa of 8.2, and the solubility greatly varies from 0.29 mg/mL at pH 7.6 to 4.4 mg/mL at pH 6.6 (44).

Preparations

Parenteral formulations include solutions, suspensions, emulsions, powders for reconstitution (solution, suspensions, liposomes), in-situ forming gels, implants, and drug-eluting devices.

Parenteral formulations, especially suspension-based formulations, should be assessed for syringeability and injectability. Syringeability and injectability are closely related. Syringeability is the ability of the parenteral formulation to easily pass through a hypodermic needle from the storage vial to the injection syringe, including ease of withdrawal, a tendency for clogging and foaming, and accuracy of dose measurements. Injectability is the performance of the formulation during injection, involving the pressure and force required for injection into the body tissues, evenness of flow, and tendency of clogging. Factors affecting syringeability and injectability include the suspension solids concentration, particle size and shape, and viscosity (50, 52). Needle clogging may occur due to blockage by a single particle or the bridging effect of multiple particles (52). The individual particle size should be no greater than one-third of the needle's internal diameter (53).

Technologies used for long-acting products include prodrugs, oil-based solutions, aqueous suspensions, polymer-based systems (e.g., microparticles), in-situ forming systems, and solid implant devices (50, 54).

Long-acting prodrugs generally involve esterification with long-chain fatty acids to increase drug lipophilicity and reduce the aqueous solubility of the active drug. Prodrugs are converted to their active component through hydrolysis (45). An example is paliperidone palmitate which hydrolyzes to the active moiety (paliperidone).

Long-acting oil-based solutions are generally lipophilic drugs or prodrugs with high log P, solubilized in an oily vehicle (e.g., arachis oil or sesame oil). Following SC or IM injection, an oily depot forms. Unlike aqueous vehicles, the rate of clearance of the oily vehicle is generally slower than the rate of drug absorption (42). Factors affecting drug release from oil-based formulations include the drug concentration, vehicle viscosity, drug solubility in the oily phase, the partition coefficient between the oily vehicle and the surrounding physiological fluid, surface area of the oily depot, and the diffusion layer thickness (40). Higher viscosity may result in poor syringeability and injectability; and slows the diffusion and dissolution rates (50).

Polymer-based systems include polymeric microparticles and implants. Polymeric microparticles contain API entrapped in a biodegradable matrix. Polymeric implants contain API entrapped in either a biodegradable or non-biodegradable polymer matrix as a cylindrical rod, generally 10-35 mm in length and 1-3 mm in diameter. Both polymeric microparticles and implants are suitable for both water-soluble and insoluble drugs, small molecules, and peptides. Drug release may be modified by varying the polymer properties (e.g., copolymer ratios, molecular weight, and end groups) (45). Microparticles may exhibit an initial burst release from surface erosion, followed by slow release from diffusion through the polymeric matrix (50).

In-situ forming systems (IFS) are liquid or semi-solid formulations that form a depot after injection. IFS was initially developed in the 1980s. The first approved in-situ forming drug depot was approved in 2002 (leuprorelin acetate, Eligard™) (55). This SC injection product is reconstituted to provide a suspension that forms a solid drug depot following injection. There are various mechanisms for depot formation, including solvent extraction, pH or temperature-induced conversion, or cross-linking polymer or monomer chains (45).

Solid implants include matrix and reservoir systems. They may contain biodegradable or non-biodegradable polymers. Drug release generally occurs by diffusion through the polymer matrix or polymer degradation. The release kinetics may be tailored by implant size, polymer composition, and drug loading. Solid implants are generally manufactured by hot melt extrusion or injection molding (45).

Long-acting aqueous suspensions contain poorly soluble drug particles suspended in an aqueous vehicle. The drug particles may be in the microparticle or nanoparticle size range. Aqueous injectable suspensions have been approved since the 1950s. Early microparticle suspension products include penicillin G benzathine and depot medroxyprogesterone acetate (DPMA). The first nanoparticle suspension product was approved in 2009, a one-monthly IM injection of paliperidone palmitate (Invega Sustena™) that is manufactured by high-energy ball milling technology (45). Aqueous suspension formulations generally contain surfactants and/or polymers used to stabilize the suspension by reduction of interfacial surface tension, steric hindrance, or electrostatic repulsion of charged particles (54). Other excipients may include viscosity enhancers, wetting agents, buffers, and preservatives (50). Aqueous suspension formulations allow for higher drug loading (up to 1000 mg) compared to polymeric microparticles and liposomes. This formulation type is suitable for poorly aqueous soluble drugs or prodrugs (45). The solid crystalline particles are less prone to chemical degradation, hydrolysis and oxidation, and polymorphic transitions (44).

RESPIRATORY AND NASAL DELIVERY

Biology

The respiratory tract is characterized by two distinct regions for drug delivery. The upper respiratory tract consists of the nasal passages in which inhaled air is heated and humidified, and ambient airborne materials, solid particles or liquid droplets (>5 μm), are collected. The nasal passage consists of nasal folds at the surface of which is a ciliated epithelium on which a layer of mucus resides (56, 57). The epithelium consists of a variety of cells including those involved in immunological response (M-cells and dendritic cells) (58). The lower respiratory tract is entered through the larynx, to the trachea, and subsequently through branching bronchi and conducting bronchioles to the periphery of respiratory bronchioles and alveoli (59) (Figure 9.3). The branching airways of the lungs are considered to be at least 23 generations (60). The upper airways, approximately 16 generations, consist of ciliated epithelium on which a mucus layer exists. The peripheral airways, approximately 17–23 airway generations, include alveolar epithelial cells with a flattened morphology that give rise to a large surface area of approximately 100 m^2, varying according to phenotypic factors dictating individual size, such as age and gender. The airways also act as a filter for incoming air, collecting ambient airborne materials >5 μm in particle diameter (61). The periphery of the lungs is the site of gaseous exchange. Any material deposited in the periphery of the lungs is cleared by cell-mediated transport involving phagocytosis by macrophages (59).

Biopharmaceutics

The disposition of particles and their component drug from the nose and lungs is dictated by the physicochemical properties of the drug particle and the mechanisms of clearance from the airways (62). Drugs that are highly soluble and delivered in particles that dissolve rapidly are subject to absorption from their site of deposition (63, 64). In the periphery of the lungs, the barrier to penetration from the airways to the systemic circulation is small and consists of a surfactant layer, and a thin epithelium composed of alveolar type 1 cells that are proximal to the capillaries of the systemic circulation (59). In the airways and nasal passages, the barriers are more substantial, including a mucus layer, ciliated columnar epithelial cells, and interstitium in which blood vessels carrying the nasal and bronchial blood supply are transported (65, 66).

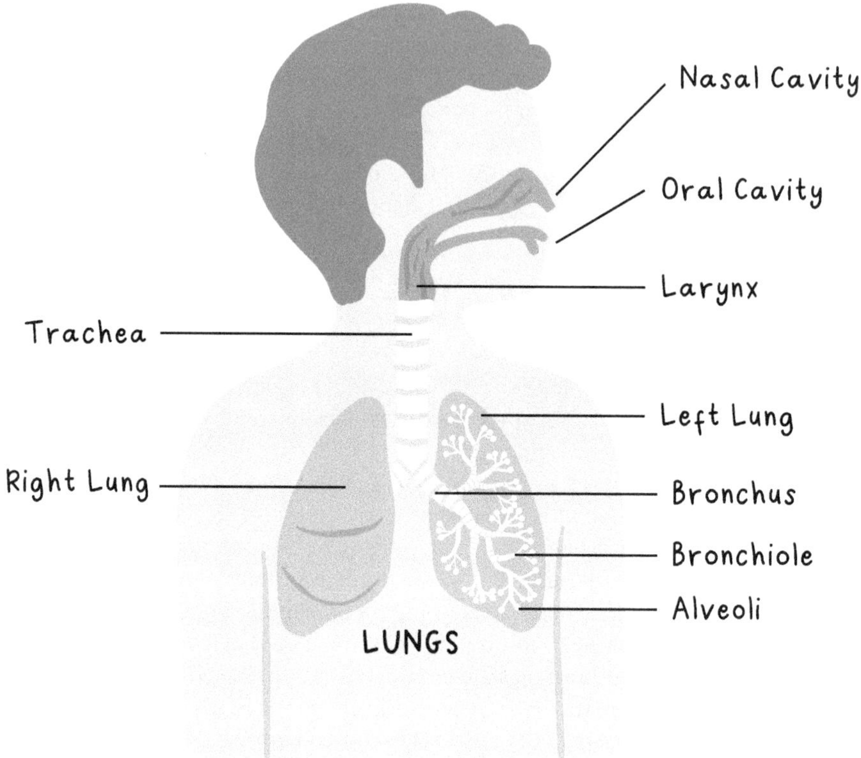

FIGURE 9.3 The respiratory tract (https://www.shutterstock.com/image-vector/human-respiratory-system-infographic-kids-study-2217855411).

Particles that have limited solubility and extended dissolution rates are subject to other mechanisms of clearance (62, 67). In the periphery, particles that are resident for extended periods of time are phagocytosed by alveolar macrophages which can act as a depot for drug release if dissolution continues within the cell or can translocate the particles to the lymphatics if dissolution is extremely delayed. Slowly dissolving particles that deposit on the mucociliary escalator, either in the nose or the airways, are subject to the combination of movement under the influence of cilia at the surface of the epithelium and dissolution accompanied by diffusion to the epithelium where absorption can occur.

It should be noted that unlike other routes of administration particles delivered to the nose or lungs predominantly deliver drugs intended for local action. There are so few exceptions to this observation that they will not be discussed here. Consequently, the biopharmaceutics of drug delivery to the lungs considers disposition from the perspective of the amount of drug remaining in the lungs rather than appearing in the systemic circulation. Conventional pharmacokinetic parameters associated with the appearance of drug in plasma are used as indicators of residence time in the lungs rather than corollaries of circulating concentrations available to the site of action.

An inhaled Biopharmaceutical Classification System (iBCS) has been proposed following the model established for the gastro-intestinal BCS (gi-BCS) used to address the suitability of drugs for oral administration. The same principles have been adopted of a grid defining four classes of drugs based on low and high solubility and permeability (68). It is noted that unlike the GI tract which may be considered a unidirectional tube with certain important physiological characteristics, notably pH, the lungs consist of a bidirectional tube and an enclosed terminus (bucket) into which drugs are introduced on an airstream and then removed through dissolution. A model that incorporates features of the lung anatomy and physiology and the physicochemical properties of the drugs has been employed to establish the specific values of the grid lines that distinguish low and high solubility and permeability (69). Finally, commercial drugs have been placed on the grid based on their known solubility and permeability and the model seems to accurately predict the grid values with respect to the placement of these products (70). It is important to recognize that the implications for the safety and efficacy of the giBCS and iBCS differ. The gi-BCS predicts the availability of drug for absorption where the pharmacological effects are largely accessed through the systemic circulation and are at remote sites. The site of action of most inhaled drugs is in the airways or lung tissue. Consequently, the iBCS is important for its prediction of local availability and residence time. The appearance of drug in the systemic circulation is a reflection of drug remaining in the airways but is influenced by factors other than solubility and permeability, such as tissue binding and lysosomal storage, that may not be significant for local drug availability.

Preparations

The approaches to the preparation of particles for the delivery of drugs to the nose and lungs are limited to methods that can produce respirable particles. In this regard, the dominant methods have been jet milling and spray drying (71, 72). Other methods have been evaluated but none have made it to commercialization. The most prominent of these alternative methods was supercritical fluid manufacture (73, 74).

Jet milling of particles produces largely crystalline material. Control of the surface amorphous content resulting from the mechanical energy expended in reducing the starting crystal structure to particles <5 μm is important for ensuring the stability of the final product. The presence of amorphous material can lead to the uptake of moisture, recrystallization, and solid bridging, all of which influence the physical stability and performance of the final product.

Milled particles have been a central component of pressurized metered dose and dry powder inhaler (pMDI and DPI) products for 70 years (75, 76). The original pMDI products involved the suspension of milled particles in a propellant medium that helped meter the dose and also disperse the drug into the air for the patient to inhale. The suspension was formulated as a flocculation with the use of surfactant such that on shaking the particles dispersed uniformly and when actuation the equilibration of the propellant vapor pressure with atmosphere resulted in a rapid evaporation and propulsion leading to high shear delivery and presentation of particles in respirable sizes.

It is generally believed that the strength of the interparticulate forces in milled powders was too strong to be overcome by the shear forces under the influence of the patient's inspiratory flow from a dry powder inhaler. Consequently, large lactose carrier particles were employed to both dilute potent low-dose drugs to facilitate dose filling and to promote dispersion into the patient's inspiratory flow (77, 78). The stability and performance of this complex formulation were dictated by the structure and size of the alpha-lactose monohydrate carrier particles, as well as drug particles and both cohesive and adhesive forces of interaction. It has recently been demonstrated that some milled particles disperse at the low shear forces of dry powder inhalers due to their unique physicochemical properties (79, 80). This may result in opportunities for higher-dose delivery of milled powders from DPIs.

Spray drying of inhaled drugs was first proposed in the late 1990s (81). Spray drying usually results in amorphous particles that have a unique structure. Notably, the tendency to form hollow, collapsed hollow, or perforated spheres results in low-density particles. These structures result in large geometric sizes but small aerodynamic sizes which in turn reduce the interparticulate forces through, few particle contacts, reduced van der Waals forces and greater interaction with the dispersing air.

The application of spray drying to pMDI products has been through the inclusion of lipid-based spray-dried excipient particles with associated milled drug (82, 83). The spray-dried excipient particle allow uniform dispersion of the milled particles in propellant and assure the dose uniformity, stability, and reproducible performance of the product.

The dry powder application of spray-dried material has focused on the ease of dispersion of particles prepared by this method. Large doses can be delivered reproducibly without the need for a carrier lactose particle. This allows for new categories of drugs to be delivered that are low in potency and have a high dose requirement. One notable example is tobramycin DPI used to treat *Pseudomonas aeruginosa* infection (84).

Supercritical fluid manufacture of particles for pharmaceutical applications also emerged in the late 1990s (73, 74). The use of liquid carbon dioxide as a means of creating particle from a molecular dispersion by controlled flash outgassing allowed for the preparation of respirable particles. The control of particle size distribution and the absence of mechanical impacts, as required for jet milling, resulted in a crystalline drug with more uniform surface properties that resulted in more predictable particle interactions. This technique was employed for the preparation of hydroxyergotamine particles for inclusion in a pMDI product for the treatment of migraine headaches. The product received regulatory approval but was not commercialized.

TRANSDERMAL AND TOPICAL DELIVERY

The topical route of administration delivers drugs to the skin tissues for local activity, whereas the transdermal route delivers drugs through the skin for systemic activity (85). Transdermal and topical drug delivery is convenient, allows self-administration, and is generally non-invasive. Transdermal drug delivery avoids the acid/base chemical degradation, enzymatic metabolism, and hepatic first-pass effect associated with the oral route of administration. It eliminates the risks and inconveniences of parenteral administration and improves patient compliance. However, drugs for transdermal delivery must have high potency and penetrate the skin barrier without being affected by the enzymes in the epidermis (15).

Biology

The skin is the largest organ of the human body constituting about 15% of the human body. The primary functions of the skin include containment of tissues and organs, protection from dangers (micro-bacterial, chemical, radiation, thermal, and electrical), environmental sensing (tactile/pressure, pain, and thermal), metabolism, disposal of biochemical waste through secretions, and blood temperature regulation. The surface area of the skin is approximately 2 m^2. The pH of the skin ranges between 4.8 and 6.0 (85–90).

The skin consists of three main layers: epidermis, dermis, and hypodermis (Figure 9.4). It is also embedded with appendages, including hair follicles, sebaceous glands, and sweat glands. Sebum is excreted by the sebaceous glands into the hair follicles. Sebum is a lipid mixture containing glycerides, waxes, squalene, fatty acids, cholesterol, and its derivatives. It contributes to the skin pH (91).

The epidermis is the outermost layer of the skin. Its thickness varies across anatomical regions, with the palms of the hand and soles of the feet being the thickest and the eyelids being the thinnest. The epidermis consists of the stratum corneum, stratum lucidum, stratum granulosum, stratum spinosum, and stratum germinativum (basale). The stratum corneum is the primary barrier for transdermal drug penetration. It is the outermost layer of the epidermis and comprises 10–15 layers of nonviable keratinocytes (corneocytes) and dendritic cells with intercellular lipid phase, containing long chain ceramides, free sterols, free fatty acids, triglycerides,

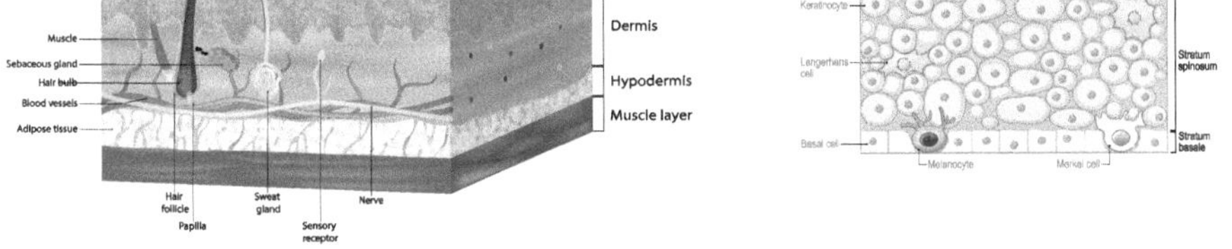

FIGURE 9.4 (A) The structure of the skin (https://www.shutterstock.com/image-vector/structure-human-skin-anatomy-detailed-vector-2341813273) and (B) the epidermis (https://www.shutterstock.com/image-vector/layers-epidermis-epithelial-cells-keratinocytes-melanocyte-2104655831).

sterol esters, and cholesterol sulfate, arranged in a multi-layered format. It is typically described as a "brick and mortar" structure with corneocytes (bricks) inter-dispersed in a lipid intercellular matrix (mortar). The thickness of the stratum corneum is approximately 10–15 μm in the dry state and up to 40 μm in the hydrated state. The corneocytes are constantly shed and replaced by differentiating keratinocytes migrating from the viable epidermis. The stratum corneum and viable epidermis possess minimal vascularization (85, 86, 89, 90, 92, 93).

The dermis is the next layer and has a thickness of 2-3 mm. It consists of a loose connective tissue matrix which provides strength and pliability. Fibroblasts, macrophages, mast cells, leukocytes, and other cells are scattered throughout the dermis. The dermis is highly vascularized and contains lymphatic vessels and nerve endings. The blood flow rate is about 0.05 $mL/min/mm^3$ and is altered by physiological and psychological responses, such as exercise, temperature, and shock. The dermis contains many enzymes, including cytochrome P450 (86, 89, 92, 93).

The hypodermis is the innermost layer of the skin and mostly consists of adipose tissue. It serves as a connecting lamina for the skin and inner tissues, muscles, and bones. The hypodermis protects against mechanical and thermal damage (86).

Biopharmaceutics

Transdermal drug delivery occurs via two pathways: transepidermal and transappendageal. The transepidermal pathway involves drug penetration through the stratum corneum. The stratum corneum provides the rate-limiting barrier to transdermal drug absorption. Drug permeability is faster through the viable epidermis, dermis, and capillary walls. Drug penetration through the stratum corneum occurs by the intracellular or intercellular pathways. The intracellular pathway involves drug transport through the corneocytes, keratinocytes, and joining lipid matrix between layers. It involves repeated partitioning and diffusion across the multiple layers of the stratum corneum. Hydrophilic and polar molecules are generally transported using the intracellular pathway and may involve passive diffusion and/or active transport mechanisms. The intercellular pathway involves drug diffusion through the lipid matrix and is favored by lipophilic molecules. Passive diffusion is governed by Fick's first law of diffusion, in which the concentration gradient is the driving force for diffusion (see Chapter 7). Transdermal drug absorption is slower and more selective than gastrointestinal absorption (15, 85, 86, 89, 90, 93).

The transappendageal pathway involves drug transport via the hair follicles and sweat glands. This pathway is mostly for water-soluble molecules. The drug may accumulate in the appendages and act as a drug reservoir. There is limited drug transport using the transappendageal pathway as appendages occupy a very low surface area of the total skin surface (0.1%) (85, 87, 89, 90). Microparticles (3–10 μm) may enter the hair follicle orifices. Nanoparticles with sizes of 300–500 nm may penetrate to the sebaceous glands (91).

Factors affecting transdermal drug delivery include the API physicochemical properties, the dosage form and formulation properties, and the physiological conditions of the skin tissue. The API physicochemical properties that affect transdermal absorption include potency, solubility, lipophilicity, polarity, octanol-water coefficient (log P), pK_a, degree of ionization, crystallinity, and molecular weight. Highly lipophilic drugs penetrate the skin easily. Optimal physicochemical drug properties include log P values between 1 and 4, low melting point below 200 degrees C, high potency with a daily dose of less than 20 mg per day, and molecular weight lower than 500 Da. A prodrug approach may be utilized to increase transdermal drug permeability, where a chemical moiety is attached to the API to improve its hydrophobicity or lipophilicity. Drugs should not produce local irritation or sensitization in the skin tissues (41, 85, 86, 89, 90, 93).

Physiological factors affecting transdermal absorption include anatomical location, degree of skin hydration, microcirculation, surface temperature, and surface sebum film. The structure of the skin is affected by diet and nutrition, sun exposure, skincare habits, racial and gender differences, age, and disease states. Some skin regions provide greater drug permeability, such as the genitalia, face, and scalp. The stratum corneum composition is altered in some disease states (e.g., psoriasis and atopic dermatitis) which results in potential greater drug penetration. The skin becomes thinner with lower lipid levels with aging. Drug penetration improves with skin hydration and warmer skin temperatures. The skin pH may influence drug penetration of weakly acidic or basic compounds (85, 86, 93, 94).

Drug absorption from transdermal delivery may be enhanced by chemical, biochemical, and physical methods. Chemical penetration enhancers improve drug partitioning and diffusion by altering lipid composition of the stratum corneum, fluidizing the lipid bilayer or altering the keratin conformation, and disrupting the cohesion of the corneocytes. Biochemical penetration enhancers disturb the stratum corneum lipid bilayer via lipid metabolism. Toxicity, irritancy, and related safety concerns are associated with chemical and biochemical penetration enhancers (87, 93).

Physical methods include iontophoresis, electroporation, sonophoresis, thermal or laser ablation, needle-free jet injections, and microneedles (85, 86, 89, 90, 92, 93, 95).

Preparations

Traditional preparations for topical and transdermal drug delivery include creams, ointments, pastes, lotions, gels, and patches.

Nanosystems have been investigated for transdermal drug delivery systems, due to their small size and potential to penetrate through the skin or act as penetration enhancers. The different types of nanoparticles include: (1) lipid-based nanosystems, such as liposomes, deformable liposomes, lipid nanoparticles (LNPs), and nanostructured lipid carriers (NLCs); (2) polymer-based nanosystems, such as micelles, polymersomes, dendrimers, nanoparticles, nanocarriers, and nanogels; (c) nanoemulsions; and (d) inorganic nanosystems, containing carbon or metals.

Liposomes are liquid spherical vesicles consisting of a lipid bilayer and an aqueous core. The lipid bilayer consists of phospholipids and cholesterol. They allow the entrapment of both hydrophilic and lipophilic drugs. Liposomal formulations have been used for drug delivery since the 1980s. A second generation of liposomes is known as deformable liposomes (e.g., ethosomes, niosomes, invasomes, and transferosomes). The liposome size, lamellarity, surface charge, coating, and aggregation may affect the degree of drug penetration. Deformable liposomes are believed to achieve deeper stratum corneum as they can squeeze through microscopic channels. Pharmaceutical transdermal products utilizing liposome and deformable liposome technology are commercially available (87).

Lipid nanoparticles are a colloidal dispersion composed of a dispersed lipid phase stabilized by an emulsifier. They include solid lipid nanoparticles (SLN)s and nanolipid carriers (NLCs) (91). SLNs are composed of solid lipids (e.g., glycerides and waxes), whereas NLCs are composed of a lipid matrix of solid and liquid lipids. These lipids are generally regarded as safe (GRAS) and non-irritating. Enhanced drug penetration of SLNs and NLCs is believed to be due to the formation of a lipid layer that adheres to the skin surface resulting in increased skin hydration. NLCs exhibit higher solubilization and deeper penetration compared to SLNs. Physicochemical factors affecting transdermal penetration include particle size, crystallinity, lipophilicity, and surface modifications. Deeper penetration was observed with smaller particle sizes and amorphous particles, which leads to faster dissolution. Pharmaceutical transdermal products utilizing lipid nanoparticle technology are commercially available (87, 91, 93).

Biodegradable polymeric nanoparticles are currently being investigated for transdermal drug delivery. Polymeric nanoparticles may contain natural polymers (e.g., chitosan and alginates) or synthetic polymers (e.g., PLA, PLGA, and PCL). Drugs may be encapsulated in the particle core, dispersed within the particle matrix, or adsorbed onto the particle surface. Generally, the polymeric nanocarrier is deposited into the stratum corneum, and the drug is released and penetrates the deeper layers (87, 96). Many studies have found that polymeric nanoparticles are unable to penetrate the stratum corneum but utilize the transappendageal/follicular pathway. The follicular transport is dependent on the nanoparticulate composition, size, shape, and other factors.

The combined use of polymeric nanoparticles with microneedle technology (see below) has been investigated to increase drug penetration to the viable epidermis. Significantly enhanced permeation of nanoparticles is observed with microneedles, compared to treatments without microneedles. Greater nanoparticle permeation occurred with smaller particle sizes and larger pore sizes. However, the maximum dimensions of the microneedles and subsequent microchannels are restricted by concerns regarding limiting invasiveness, pain, and safety concerns. Greater hydrophilicity allows deeper penetration into the aqueous environment of the microchannels. Negatively charged nanoparticles showed greater permeability, compared with positively charged nanoparticles (97).

Inorganic nanoparticles containing carbon, silver, and gold have been investigated as transdermal drug-carriers as nanotubes, nanorods, and nanospheres. However, they have limited use due to safety and biocompatibility issues (87).

Microneedle technology has been evaluated as a physical method to enhance drug penetration of microparticles, nanoparticles, and nanosuspensions. Microneedle-assisted nanoparticles and nanosuspensions have shown greater transdermal penetration compared with needle-free delivery systems, attributed to disruption of the stratum corneum barrier. Microneedles consist of needle-like micro-projections positioned on a baseplate. The microneedles penetrate through the stratum corneum into the viable epidermis and dermis and create transient aqueous microchannels. The needle length is optimized to avoid stimulation of dermal nerve endings. They are minimally invasive and generally painless. There are five types of microneedles, including: solid, coated, hollow, dissolving, and hydrogel-forming. Solid microneedles are inserted to create a transient pore, then removed, and a drug-loaded reservoir is applied for drug delivery. Coated microneedles have a drug coated on the outer surface of a solid microneedle, where the drug remains deposited in the dermal tissue after removal of the microneedle. Hollow microneedles contain a central channel that allows drug delivery to the dermal tissue. Dissolving microneedles contain API dispersed in a matrix of biodegradable polymers or sugars [e.g., polyvinyl alcohol (PVA), polyvinylpyrrolidone (PVP), and carboxymethyl cellulose (CMC)]. These microneedles are inserted into the skin, and drug is released by dissolution into the interstitial fluid of the dermal tissue. Hydrogel-forming microneedles contain a drug reservoir loaded on top of the microneedle array. These microneedles are inserted and rapidly swell with absorption of the interstitial fluid. Drug delivery is controlled by diffusion through the swollen hydrogel and residence time that the microneedle is left in the dermal tissue. Preclinical and clinical studies have shown some promising results. Work is still needed, including the commercial-scale manufacturing (95, 98).

BUCCAL AND SUBLINGUAL DELIVERY

The sublingual and buccal mucosal surfaces of the oral cavity provide an opportunity for systemic drug delivery. The sublingual route of administration involves placing the drug under the tongue and the buccal route of administration involves placing the drug between the gums and cheek. Advantages of sublingual and buccal delivery include neutral pH and lower enzymatic activity (compared with other parts of the gastrointestinal tract), faster onset of action (compared with oral administration), and avoidance of the hepatic first-pass metabolism. The limitations include the difficulty of maintaining dosage forms at the sublingual or buccal sites, the small surface area available for drug absorption, the need for high-potency drugs for small doses, the potential for drugs to be unpalatable or cause irritation, the risk of accidental aspiration, and the lack of suitability for unconscious or uncooperative patients (99, 100).

Biology

The oral cavity is lined with nonkeratinized, stratified squamous epithelium cells covered with a mucus layer thickness of 70–100 μm. Saliva lubricates and protects the oral mucosa and has a relatively neutral pH (6.2–7.4) with low enzymatic activity. The saliva hydrates the oral mucosa surface and provides a small fluid volume for drug dissolution. The sublingual region has a surface area of 26.5 ± 4.2 cm^2 with an epithelium layer thickness of 8–12 cells (100–200 μm). The buccal region has a surface area of 50.2 ± 2.9 cm^2 with an epithelium layer thickness of 40–50 cells (500–800 μm). The cell turnover for the sublingual and buccal regions is 20 days and 13 days, respectively. Faster absorption and onset of action are observed for sublingual administration due to the thinner epithelium. The mucosal tissue beneath the epithelium layer contains a network of blood capillaries and lymph vessels. The venous blood drains into the superior vena cava, bypassing the hepatic first-pass metabolism. (92, 99, 100).

Biopharmaceutics

Drug absorption across the oral epithelium occurs mainly by passive diffusion via transcellular and paracellular routes. Drug absorption is affected by API physicochemical properties, including lipid solubility, partition coefficient, molecular weight, solubility at the site of absorption, and degree of ionization (2, 99).

Physiological factors affecting sublingual and buccal absorption include the residence time, saliva flow rate and pH, and the mucosal membrane integrity. The drug is released from the dosage form and dissolves in the saliva before absorption occurs. The extent of drug absorption is reduced if the formulation or dissolved drug is swallowed before absorption occurs. Longer residence time may be observed with specific preparation types (e.g., gels, films, and patches). The saliva provides the fluid medium for drug dissolution. There may be insufficient volume for drug dissolution if the saliva flow is low and the mouth is dry. However, dissolved drug may be swallowed before absorption occurs if the saliva flow rate is too high. The saliva flow rate may be affected by various factors. Low saliva flow rate is observed with dehydration, age, anticholinergic drugs, stress or anxiety, disease states, smoking, and physical activity. A high saliva flow rate is observed with chewing; anticipation of food; and acidic, sour, or spicy foods. The pH of the saliva is typically neutral but may be temporarily altered by food, drinks, or oral diseases. The drug absorption is affected if the oral mucosal membrane is inflamed or has open sores (99, 100).

Preparations

Preparations for buccal or sublingual delivery include tablets, lozenges, films, gums, gels, liquids, sprays, and aerosols. Development strategies to improve sublingual and buccal drug absorption include increasing mucosal retention (e.g., mucoadhesive polymers) and/or permeation enhancers (e.g., surfactants, bile salts, fatty acids, amino acids, and cyclodextrins). Novel technologies include nanoparticles, extruded films, and printed films (100, 101).

RECTAL DELIVERY

The rectal route of drug administration provides another alternative to oral administration and may be used for local and systemic action. It provides systemic drug delivery for patients where oral administration is unsuitable, for example, patients with nausea, vomiting, unconscious, or swallowing difficulties. It also provides local drug delivery for rectal conditions (e.g., constipation and hemorrhoids). The advantages of rectal delivery include self-administration, rapid absorption, and avoidance of gastrointestinal irritation. The rectum has lower enzymatic activity than other areas of the gastrointestinal tract. Drugs absorbed in the lower portion of the rectum avoid metabolic first-pass metabolism. The limitations include the limited surface area for drug absorption, limited fluid volume for drug dissolution, and lack of patient preference due to cultural concerns and/or discomfort and leakage. Preparations for rectal delivery include suppositories, enemas, foams, ointments, and gels (15, 102, 103).

Biology

The rectum is the terminal segment of the large intestine. The primary function is a conduit and storage site of fecal matter prior to defecation. The rectum in adults is 15–20 cm in length, with a surface area of 200–400 cm^2. The rectal fluid is approximately 1–3 mL, with a pH of 7–8 and minimal buffering capacity or enzymatic activity. Some residual bacterial enzymes are present in the rectum which originates from the colonic microbiome. The mucosal surface of the rectum consists of a single layer of columnar cells with some interspersed goblet cells. The rectal epithelium transitions from non-keratinized cells at the anorectal junction to keratinized cells at the external anus sphincter. The mucus secreted from the goblet cells protects and lubricates the rectal epithelium. The rectal mucus layer thickness is ~150 μm with an estimated turnover time of 3–4 hours. The rectal tissue contains a network of blood capillaries and lymph vessels. Drugs absorbed from the upper portion of the rectum may enter the superior hemorrhoidal vein which drains directly into the portal vein and is subject to hepatic first-pass metabolism. Drugs absorbed from the lower portion of the rectum may enter the inferior and middle hemorrhoidal veins which drain into the inferior vena cava and bypass the hepatic first-pass metabolism. Highly lipophilic drugs may be absorbed into the lymphatic system, which also bypasses the hepatic first-pass metabolism (15, 41, 102).

Biopharmaceutics

Drug absorption across the rectal mucosa generally occurs by passive diffusion via paracellular or transcellular pathways. Drug absorption from rectal drug delivery is determined by API physicochemical properties, formulation factors, and physiological variables. The drug must be released from the formulation, and then dissolve in the rectal fluids before absorption across the mucus layer and rectal epithelium.

The API physicochemical properties affecting drug absorption include solubility, pK_a, degree of ionization, lipophilicity, molecular size, and polarity. Non-ionized forms are more readily absorbed. An optimal balance is required between hydrophilicity and lipophilicity. Drugs should have adequate hydrophilic properties to be soluble in the rectal fluids but also be sufficiently lipophilic to cross the epithelium. Techniques to improve the solubility and dissolution rate of poorly soluble drugs include particle size reduction, salt selection, and use of surfactants (102, 104, 105).

The physiological factors affecting drug absorption include the presence of fecal material, rectal fluid pH, residence time, disease state, and concomitant drug therapy. The presence of fecal matter increases the rectal fluid viscosity, which subsequently affects drug dissolution. The rectal fluid has a low buffering capacity so the pH may be altered with use of external products, which may in turn affect drug ionization, dissolution and absorption. The residence time of the rectal dosage form depends on the frequency of defecation. This varies within and between individuals depending on their diet, disease state, diurnal, and other factors. The colonic transit time may range from 6 to 70 hours. Early expulsion of the dosage form may result in reduced drug absorption. Some disease conditions such as inflammatory bowel disease, hemorrhoids, and gastrointestinal infections may affect tissue integrity, inflammation, and gastrointestinal motility; which may affect drug absorption. Also, local tissue trauma, for example, ruptured hemorrhoids, may lead to painful drug administration into the rectal cavity (15, 102, 105).

Preparations

Rectal dosage forms may be liquid (e.g., enemas), semisolid (e.g., gels and foams), or solid formulations (e.g., suppositories). Solid dosage forms must disintegrate and dissolve before drug absorption can occur. Therefore, liquid and semi-solid dosage forms may provide faster drug absorption. Suppositories are the most common rectal dosage form. They dissolve in the rectal secretions and the drug is absorbed in the rectal mucosa. Generally, lipophilic drugs are formulated in hydrophilic bases (e.g., polyethylene glycol), and hydrophilic drugs are formulated in lipophilic bases (e.g., cocoa butter). Lipophilic bases melt readily at body temperature to release drug, whereas hydrophilic bases dissolve in the physiological fluids to release drug. Drugs may be dissolved or dispersed in the suppository base. Suppositories should be retained in the rectum for at least 20–30 minutes to ensure complete melting and dissolution of the base and drug release. Drug dissolution may be enhanced by particle size reduction or surfactants. Novel technologies investigated for rectal drug delivery for systemic and local activity include nanoparticulate systems, solid lipid nanoparticles, and mucoadhesive and/or thermosensitive polymer systems (100).

VAGINAL DELIVERY

The vaginal route of administration may be used for local action (e.g., antimicrobials, antifungals, antivirals, and spermicidal agents) and systemic action (e.g., contraception and hormonal replacement therapy). The advantages of vaginal administration include good permeability, high vascularization, ease of self-administration, discreet use, avoidance of gastrointestinal degradation, and avoidance of hepatic first-pass metabolism. It may be useful for drugs with poor oral bioavailability or those that produce gastric irritation. Limitations include gender specificity, the influence of sexual activity, variability in drug absorption due to physiological conditions (e.g., age, hormone levels, and menstrual cycle), and individual beliefs based on personal hygiene and cultural sensitivity (15, 106–109).

Biology

The vagina is a slightly S-shaped fibromuscular structure that extends from the vulva to the cervix, approximately 6–12 cm in length. The vagina is positioned between the urethra, bladder, and rectum (108). The vaginal histology consists of four distinct layers. The first superficial layer is the tunica mucosa consisting mainly of nonsecretory, stratified epithelium. The second layer is the lamina propria made of collagen and elastin and contains the vascular and lymphatic vessels. The third layer is the tunica muscularis composed of smooth muscle fibers running in both circular and longitudinal directions. The final layer is the tunica adventitia consisting of areolar connective tissue. The inner surface of the vagina contains transverse folds, known as rugae, which provide distensibility, support, and increased surface area (106, 108 –110).

The vagina is highly vascularized, with supply from the internal iliac, uterine, pudendal, and middle and inferior hemorrhoidal arteries. The vaginal veins form a venous plexus and drain direct into the inferior vena cava and bypass the hepatic first-pass metabolism (106–108). The vaginal secretions are mostly transudate from vaginal and cervical cells. The vaginal fluid also contains vulvar secretions; cervical mucus, endometrial, and oviductal fluids; and microorganisms and their metabolic products (106). The vaginal epithelium is coated by a mucus layer, which contains mostly water (95%) with 2–3 % solid matter. The cervicovaginal mucus consists of mucin glycoproteins, carbohydrates, fatty acids, proteolytic enzymes, lactic acid, and acetic acid. Mucins glycoproteins (10^6–10^7 kDa) have a "bottle brush-like" structure with hydrophilic and hydrophobic regions. The mucin glycoproteins interact and form a 3D mesh network containing aqueous channels of 50–1800 nm diameter. The cervicovaginal mucus is rapidly eliminated through mucus turnover and discharge. Its main function is protection against vaginal infections and for lubrication (111 –115). Although there is relatively low enzymatic activity in the vagina, compared with the gastrointestinal tract, there is a wide range of enzymes present, including nucleases, lysozymes, and esterases (109, 111). Lactobacillus are the most prevalent microorganism in the vaginal microbiota in women of reproductive age and produce lactic acid from glycogen metabolism. The lactic acid acts as a buffer to maintain pH levels between 3.8 and 4.2. The pH levels in prepuberal and menopausal females are slightly higher (up to pH 7) due to lower hormonal levels. Menstrual, cervical, and uterine secretions and semen are alkaline which increases vaginal pH (111).

Biopharmaceutics

Like other routes of administration, drug absorption occurs in two main steps: drug dissolution in the vaginal fluids and drug transport across the vaginal epithelium. Drug transport across the vaginal epithelium occurs by three mechanisms: (1) transcellular transport via concentration-dependent diffusion, (2) paracellular mediated by tight junctions, and (3) vesicular or receptor-mediated active transport (110).

Drug absorption from the vaginal route of administration is affected by API physicochemical properties, formulation properties, and physiological factors.

The physiological conditions affecting vaginal drug absorption include the volume, composition, and pH of the vaginal fluid; and the thickness of the vaginal epithelium. Both vaginal fluid and vaginal epithelium characteristics are influenced by physiological changes due to age, hormone levels during the menstrual cycle, pregnancy, sexual arousal, concomitant diseases, and microbial infections (15, 106, 116).

The properties of the vaginal fluid may impact drug dissolution and, subsequently, drug absorption. The volume of fluid present in the vaginal cavity is relatively small (0.5 – 1 mL) (109). The volume and composition of vaginal fluid is affected by age, hormone levels, and the state of sexual arousal. The rate of vaginal fluid production is approximately 6 g/day in women of reproductive age and is reduced by 50% following menopause (112). During sexual arousal, the vagina becomes engorged and vasoactive peptides are released which increases vasodilation and suppresses venous return, resulting in increased vaginal lubrication (106). The dissolution of poorly soluble drugs may improve with increased vaginal fluid volume; however, this may reduce the residence time in the vaginal cavity available for drug absorption. The pH of the vaginal fluid may impact the degree of drug ionization and subsequently affect the dissolution profile (110). The mucus production varies with the hormonal changes during the menstrual cycle. Estrogen predominates in the first portion of the menstrual cycle, and the mucus is mostly clear, thin, and alkaline. Progesterone dominates in the second half of the cycle, and the mucus is produced in lower amounts with higher viscosity (111–113, 115).

The vaginal mucosa has similar permeability to the buccal mucosa, which is greater than the intestinal and rectal mucosa. Drugs with molecular weights >300 kDa may be able to penetrate the vaginal mucosa (109). The thickness of vaginal epithelium varies from 0.2 to 0.5 mm depending on the hormonal levels (104, 117). The epithelium sheds constantly with the vaginal cell turnover estimated at 10–15 layers every 7 days (110). Estrogen induces epithelial growth and cornification during the proliferative phase of the menstrual cycle. The epithelial layer becomes thin during pregnancy (109) and becomes thin and atrophic during menopause (112). The cervicovaginal mucus layer forms an aqueous diffusion layer adjacent to the vaginal epithelium. The mucus diffusion layer is the rate-limiting step of absorption for drugs with high permeability, whereas the epithelial layer is the rate-limiting step of absorption for drugs with low permeability (116).

Preparations

Vaginal dosage forms include semi-solid and solid dosage forms. Traditional semi-solid preparations include creams, ointments, gels, and emulsions. Traditional solid preparations include vaginal tablets, suppositories, and pessaries that dissolve or melt inside the vaginal cavity. Some disadvantages of traditional products have been reported as messy, uncomfortable feeling, leakage, local irritation, and short retention time. These may lead to poor patient acceptability and adherence.

Novel vaginal dosage forms include vaginal rings, vaginal films, and mucoadhesive products. Vaginal rings are small, flexible, polymeric devices (generally 54 mm diameter and 4 mm thick) that are inserted in the vagina and placed near the cervix for sustained drug delivery. Vaginal films are a solid dosage form that rapidly dissolves in vaginal fluids to form a viscous mucoadhesive gel (118, 119). Mucoadhesive formulations are designed to increase residence time on the vaginal mucosa and prolong drug delivery through the formation of physical and chemical interactions with the cervicovaginal mucus. The mucoadhesive interactions occur in three stages: (1) wetting and swelling of the polymer, (2) connection of the polymer chains with the mucin chains in the cervicovaginal mucus, and (3) physical or chemical interactions via physical entanglement, hydrogen bonds, van der Waals, electrostatic interactions, or covalent bonds. Generally, stronger mucoadhesive interactions result in longer residence time of the drug formulation in the vaginal cavity. Mucoadhesive products include hydrogels, emulsions, tablets, pessaries, and microparticulate and nanoparticulate systems. Multiparticulate systems provide improved spreading across the vagina cavity. Particles outside the submicron range adhere to the top layers of the mucus due to steric impedance, whereas nanoparticles can penetrate into the smaller channels of the mucus mesh network (112, 114). Mouse model studies show that nanoparticles (100–300 nm) can penetrate the vaginal epithelium and reach the lamina propria. Nanoparticles may also migrate upwards to the uterus and Fallopian tubes (114).

CONCLUSION

The use of particles in pharmaceutical dosage forms must first consider the route of administration to be employed. The primary routes of administration are oral delivery to the gastrointestinal tract or parenteral administration, either intravenously to the systemic circulation or to localized sites by intramuscular, subcutaneous, or intradermal delivery. Alternative routes of administration include: buccal, sublingual, nasal, pulmonary, skin (transdermal or topical), rectal, or vaginal delivery. Each of these routes places specific demands on the dosage form to achieve the desired biopharmaceutical outcome. Biopharmaceutical considerations relate to the point of first contact between the dosage form and its components and the local site of delivery. The biopharmaceutics depends on both the physicochemical characteristics of the dosage form, frequently governed by the component particles, and the biology of the site of delivery (local fluid properties and tissue uptake). The pharmacokinetics of drug disposition locally and systemically must be assessed to address the bioavailability of the drug. The components of the dosage form are characterized, monitored, and controlled to assure the quality of the product and the uniformity of drug dose delivery which is correlated with safety and efficacy. Therefore, the importance of characterizing and controlling the properties of powders and particles from which most dosage forms are prepared cannot be understated.

REFERENCES

1. Mayersohn M. Principles of drug absorption. In: Banker GS, Rhodes CT, editors. *Modern Pharmaceutics*. 3rd ed. New York, NY: Marcel Dekker; 1996. p. 21–73.
2. Chien YW. Oral drug delivery and delivery systems. In: Chien YW, editor. *Novel Drug Delivery Systems*. New York, NY: Marcel Dekker; 1992. p. 139–96.
3. Hua S. Advances in oral drug delivery for regional targeting in the gastrointestinal tract - influence of physiological, pathophysiological and pharmaceutical factors. *Frontiers in Pharmacology*. 2020;11:524.
4. Abdulkarim M, Sharma PK, Gumbleton M. Self-emulsifying drug delivery system: Mucus permeation and innovative quantification technologies. *Adv Drug Deliv Rev*. 2019;142:62–74.
5. Ensign LM. Oral drug delivery with polymeric nanoparticles: The gastrointestinal barrier. *Adv Drug Deliv Rev*. 2012;64:557–70.
6. Mudie DM, Amidon GL, Amidon GE. Physiological parameters for oral delivery and in vitro testing. *Mol Pharm*. 2010;7:1388–405.
7. dosSantos AM, Goncalves-Carvalho S, Bagliotti-Meneguin A, Sabio RM, Greniao MPD, Chorilli M. Oral delivery of micro-nanoparticulate systems based on natural polysaccharides for intestinal diseases therapy: Challenges, advances and future perspectives. *J Controlled Rel*. 2021;334:353–66.
8. Lou J, Duan H, Qin Q, Teng Z, Gan F, Zhou X, et al. Advances in oral drug delivery systems: Challenges and opportunities. *Pharmaceutics*. 2023;15:484.
9. Trevaskis NI, Charman WN, Porter CJH. Lipid-based delivery systems and intestinal lymphatic drug transport: A mechanistic update. *Adv Drug Deliv Rev*. 2008;60:702–16.
10. Amidon GL, Lennernas H, Shah VP, Crison JR. A theoretical basis for a biopharmaceutical drug classification: The correlation of in vitro drug product dissolution and in vivo bioavailability. *Pharm Res*. 1995;12:413–20.
11. Shah VP, Amidon GL. G.L. Amidon, H. Lennernas, V.P. Shah and J.R. Crison. A theoretical basis for a biopharmaceutic drug classification: The correlation of in vitro drug product dissolution and in vivo bioavailability, Pharm Res 12, 413–420, 1995 - backstory of BCS. *AAPS J*. 2014;16:894–8.
12. Tsume Y, Mudie DM, Langguth P, Amidon GE, Amidon GL. The Biopharmaceutical Classification System: Subclasses for in vivo predictive dissolution (IPD) methodology and IVIVC. *Eur J Pharm Sci*. 2014;16:152–63.
13. Hoener B-A, Benet LZ. Factors influencing drug absorption and drug availability. In: Banker GS, Rhodes CT, editors. *Modern Pharmaceutics*. 3rd ed. New York, NY: Marcel Dekker; 1996. p. 121–53.
14. Martinez MN, Amidon GL. A mechanistic approach to understanding the factors affecting drug absorption: A review of fundamentals. *J Clin Pharmacol*. 2002;42:620–43.

15. Robinson DH, Narducci WA, Ueda CT. Drug delivery and administration. In: DiPiro JT, Talbot RL, Yee GC, Matzke GR, Wells BG, Posey LM, editors. *Pharmacotherapy: A Pathophysiologic Approach*. Stamford, CT: Appleton and Lange; 1996. p. 49–76.
16. Chourasia ME, Jain SK. Pharmaceutical approaches to colon targeted drug delivery systems. *J Pharm Sci*. 2003;6:33–66.
17. Philip AK, Philip B. Colon targeted drug delivery systems: A review on primary and novel approaches. *Oman Med J*. 2010;25:79–87.
18. Lin L, Wong H. Predicting oral drug absorption: Mini review on physiologically-based pharmacokinetic models. *Pharmaceutics*. 2017;9:41.
19. Norris DA, Puri N, Sinko PJ. The effect of physical barriers and properties on the oral absorption of particulates. *Adv Drug Deliv Rev*. 1998;34:135–54.
20. Hussain N, Jaitley V, Florence AT. Recent advances in the understanding of uptake of microparticles across the gastrointestinal lymphatics. *Adv Drug Deliv Rev*. 2001;50:107–42.
21. Plapied I, Duhem N, Rieux AD, Preat V. Fate of polymeric nanocarriers for oral drug delivery. *Current Opinions in Colloid and Interface Sci*. 2011;16:228–37.
22. Beloqui A, Rieux AD, Preat V. Mechanisms of transport of polymeric and lipidic nanoparticles across the intestinal barrier. *Adv Drug Deliv Rev*. 2016;106:242–55.
23. Jani PU, Halbert GW, Langridge J, Florence AT. Nanoparticle uptake by the rat gastrointestinal mucosa: Quantitation and particle size dependency. *H Pharm Pharmacol*. 1990;42:821–6.
24. Jani PU, McCarthy DE, Florence AT. Nanosphere and microsphere uptake via Peyer's patches: Observation of the rate of uptake in the rat after a single oral dose. *Int J Pharm*. 1992;86:239–46.
25. Iyer R, Jovanovska VP, Berginc K, Jaklic M, Fabiani F, Harlacher C, et al. Amorphous solid dispersions (ASDs): The influence of material properties, manufacturing processes and analytical technologies in drug product development. *Pharmaceutics*. 2021;13:1682.
26. Nimmerfall F, Rosenhaler J. Dependence of area under the curve on proquazone particle size and in vitro dissolution rate. *J Pharm Sci*. 1980;69:605–7.
27. Mosharraf M, Nystrom C. Effect of particle size and shape on the surface specific dissolution rate of micronized practically insoluble drugs. *Int J Pharm*. 1995;122:35–47.
28. Anderberg EK, Bisrat M, Nystrom C. Physicochemical properties of drug release. Part 7: Effect of surfactant concentration and drug particle size on solubility and dissolution rate of felodipine, a sparingly soluble drug. *Int J Pharm*. 1988;47:67–77.
29. Aguiar AJ, Zelmer JE, Kinkel AW. Deaggregation behavior of a relatively insoluble substituted benzoic acid and its sodium salt. *J Pharm Sci*. 1967;56:1243–52.
30. Jindal KC, Chaudry RS, Singal AK, Gangwal SS, Khanna S. Effect of particle size on the bioavailability and dissolution rate of rifampicin. *Indian Drugs*. 1995;32:100–15.
31. Allen L, Yanchick VA, Maness DD. Dissolution rates of corticosteroids utilizing sugar glass dispersions. *J Pharm Sci*. 1977;66:494–7.
32. Stavchansky S, Gowan W. Evaluation of the bioavailability of a solid dispersion of phenytoin in polyethylene glycol 6000 and a commercial phenytoin sodium capsule in the dog. *J Pharm Sci*. 1984;73:733–6.
33. McGinity JW, Ku CT, Bodmeier R, Harris MR. Dissolution and uniformity properties of ordered mixes of micronized griseofulvin and a directly compressible excipient. *Drug Dev Ind Pharm*. 1985;11:891–900.
34. Nystrom C, Westerberg M. Use of ordered mixtures for improving the dissolution rate of low solubility compounds. *J Pharm Pharmacol*. 1986;38:161–5.
35. Gursoy RN, Benita S. Self-emulsifying drug delivery systems (SEDDS) for improved oral delivery of lipophilic drugs. *Biomed and Pharmacother*. 2004;58:173–82.
36. Jannin V, Musakhanian J, Marchaud D. Approaches for the development of solid and semi-solid lipid-based formulations. *Adv Drug Deliv Rev*. 2008;60:734–46.

37. Khadka P, Ro J, Kim H, Kim I, Kim JT, Cho JM, et al. Pharmaceutical particle technologies: An approach to improve drug solubility, dissolution, and bioavailability. *Asian J Pharm Sci.* 2014;9:304–16.
38. Adeoye O, Cabral-Marques H. Cyclodextrin nanosystems in oral drug delivery: A mini review. *Int J Pharm.* 2017;531:521–31.
39. Mendes M, Soares HT, Arnaut LG, Sousa JJ, Pais CC, Vitorino C. Can lipid nanoparticles improve intestinal absorption? *Int J Pharm.* 2016;515:69–83.
40. Kalicharan RW, Schot P, Vromans H. Fundamental understanding of drug absorption from a parenteral oil depot. *Eur J Pharm Sci.* 2016;83:19–27.
41. Oie S, Benet LZ. The effect of route of administration and distribution on drug action. In: Banker GS, Rhodes CT, editors. *Modern Pharmaceutics.* 3rd ed. 1996. p. 155–78.
42. Zuidema J, Kadir F, Titulaer HAC, Oussoren C. Release and absorption rates of intramuscularly and subcutaneously injected pharmaceuticals (II). *Int J Pharm.* 1994;105:189–207.
43. Chien YW. Parenteral drug delivery and delivery systems. In: Chien YW, editor. *Novel Drug Delivery Systems.* New York, NY: Marcel Dekker; 1992. p. 381–528.
44. Nguyen VTT, Darville N, Vermeulen A. Pharmacokinetics of long-acting aqueous nano-/microsuspensions after intramuscular administration in different animal species and humans - a review. *AAPS J.* 2022;25:4.
45. Bassand C, Villois A, Bianola L, Laue G, Ramazani F, Riebesehl B, et al. Smart design of patient-centric long-acting products: From preclinical to marketed pipeline trends and opportunities. *Expert Opinion Drug Delivery.* 2022: https://doi.org/10.1080/17425247.2022.2106213.
46. Feldman S. Physicochemical factors influencing drug absorption from the intramuscular injection site. *Bull Parenteral Drug Assoc.* 1974;28:53–63.
47. Hirano K, Yamada H. Studies on the absorption of practically water insoluble drugs following injection. Part 6. Subcutaneous absorption from aqueous suspensions in rats. *J Pharm Sci.* 1982;71:500–5.
48. Butterstein GM, Castracane VD. Effect of particle size on the prolonged action of subcutaneous danazol in male and female rats. *Fertil Steril.* 2000;74:356–8.
49. Bao Q, Wang X, Zhou Y, Wang Y, Burgess DJ. In vitro release testing method development for long-acting injectable suspensions. *Int J Pharm.* 2022;622:121840.
50. Bauer A, Berben P, Chakravarthi SS, Chattorraj S, Garg A, Gourdon B, et al. Current state and opportunities with long-acting injectables: Industry perspectives from the innovation and quality consortium "Long-Acting Injectables" Working Group. *Pharm Res.* 2023;40:1601–31.
51. Zuidema J, Pieters FAJM, Duchateau GSMJE. Release and absorption rate aspects of intramuscularly injected pharmaceuticals. *Int J Pharm.* 1988;47:1–12.
52. Floyd AG, Jain S. Injectable emulsions and suspensions. In: Lieberman HA, Rieger MM, Banker GS, editors. *Pharmaceutical Dosage Forms: Disperse Systems.* New York, NY: Marcel Dekker; 1996. p. 261–318.
53. Nash RA. Pharmaceutical suspensions. In: Lieberman HA, Rieger MM, Banker GS, editors. *Pharmaceutical Dosage Forms: Disperse Systems.* 1996. p. 1–46.
54. Holm R, Lee RW, Glassco J, DiFranco N, Bao Q, Burgess DJ, et al. Long-acting injectable aqueous suspensions-Summary from an AAPS Workshop. *AAPS J.* 2023;25:49.
55. Packhaeuser CB, Schnieders J, Oster CG, Kissel T. In situ forming parenteral drug delivery systems: An overview. *Eur J Pharm Biopharm.* 2004;58:445–55.
56. Geurkink N. Nasal anatomy, physiology, and function. *JACI.* 1983;72:123–8.
57. Freeman SC, Karp DA, Kathwaji CI. Physiology, Nasal. *StatPearls [Internet].* Treasure Island, FL: StatPearls Publishing; 2023. Available from: https://www.ncbi.nih.gov/books/NBK526086.

58. Sepahi A, Salinas I. The evolution of nasal immune systems in vertebrates. *Mol Immunol.* 2015;19:131–8.
59. Mortensen NP, Hickey AJ. Targeting inhaled therapy beyond the lungs. *Respiration.* 2014;88:353–64.
60. Weibel ER. *Morphometry of the Human Lung.* Heidelberg, Germany: Springer Berlin; 1963.
61. Patton JS, Byron PR. Inhaling medicines: Delivering drugs to the body through the lungs. *Nature Reviews Drug Discovery.* 2007;6:67–74.
62. Olsson B, Bondesson E, Borgstrom L, Edsbacker S, Eirefelt S, Ekelund K, et al. Pulmonary drug metabolism, clearance and absorption. In: Smyth HDC, Hickey AJ, editors. *Controlled Pulmonary Drug Delivery.* New York, NY: Springer; 2011. p. 21–50.
63. Byron PR. Prediction of drug residence times in regions of the human respiratory tract following aerosol inhalation. *J Pharm Sci.* 1986;75:433–8.
64. Suarez S, Hickey AJ. Pharmacokinetic and pharmacodynamic aspects of inhaled therapeutic agents. In: Martonen TB, editor. *Medical Applications of Computer Modelling: The Respiratory System.* Southampton, UK: WIT Press; 2001. p. 225–304.
65. Haasbroek-Pheiffer A, VanNiekerk S, DerKooy F, Cloete T, Steenekamp J, Hamman J. In vitro and ex vivo experimental models for evaluation of intranasal systemic drug delivery as well as direct nose-to-brain drug delivery. *Biopharmaceutics and Drug Disposition.* 2023;44:1–126.
66. Gonda I. Mathematical modeling of deposition and disposition of drugs administered via the nose. *Adv Drug Deliv Rev.* 1998;29:179–84.
67. Gonda I. Drugs administered directly into the respiratory tract: Modeling of the duration of effective drug levels. *J Pharm Sci.* 1988;77:340–6.
68. Hastedt JE, Backman P, Cabal A, Clark A, Ehrhardt C, Forbes B, et al. iBCS:1. Principles and framework of an inhalation-based biopharmaceutical classification system. *Mol Pharm.* 2022;19:2032–9.
69. Backman P, Cabal A, Clark A, Ehrhardt C, Forbes B, Hastedt J, et al. iBCS:2. Mechanistic modeling of pulmonary availability of inhaled drugs versus critical product attributes. *Mol Pharm.* 2022;19:2040–7.
70. Hastedt JE, Backman P, Cabal A, Clark A, Ehrhardt C, Forbes B, et al. iBCS:3. A biopharmaceutics classification system for orally inhaled drug products. *Mol Pharm.* 2024;21:164–72.
71. Hickey AJ. Lung deposition and clearance of pharmaceutical aerosols: What can be learned from inhalation toxicology and industrial hygiene? *Aerosol Sci Tech.* 1993;18:290–304.
72. Vehring R. Pharmaceutical particle engineering via spray drying. *Pharm Res.* 2008;25:999–1022.
73. York P. Strategies for particle design using supercritical fluid technologies. *Pharm Sci Tech Today.* 1999;2:430–40.
74. Chang RYK, Chan K-K. Advancements in particle engineering for inhalation delivery of small molecules and biotherapeutics. *Pharm Res.* 2022;39:3047–61.
75. Stein SW, Thiel CG. The history of therapeutic aerosols: A chronological Review. *J Aerosol Med Pulm Drug Deliv.* 2017;30:20–41.
76. Telko MJ, Hickey AJ. Dry powder inhaler formulation. *Respiratory Care.* 2005;50:1209–27.
77. Hickey AJ, Mansour HM, Telko MJ, Xu Z, Smyth HDC, Mulder T, et al. Physical characterization of component particles included in dry powder inhalers. II. Dynamic characteristics. *J Pharm Sci.* 2007;96:1302–19.
78. Hickey AJ, Mansour HM, Telko MJ, Zhu Z, Smyth HDC, Mulder T, et al. Physical characterization of component particles included in dry powder inhalers: I. Strategy review and static characteristics. *J Pharm Sci.* 2007;96:1282–301.

79. Brunaugh AD, Jan SU, Ferrati S, Smyth HDC. Excipient-free pulmonary delivery and macrophage targeting of clofazimine via air jet micronization. *Mol Pharm.* 2017;14:4019–31.
80. Nair VV, Smyth HDC. Inhalable excipient-free dry powder of tigecycline for the treatment of pulmonary infections. *Mol Pharm.* 2023;20:4640–53.
81. Edwards DA, Hanes J, Caponetti G, Hrkach J, Ben-Jebria A, Eskew ML, et al. Large porous particles for pulmonary drug delivery. *Science.* 1997;276:1868–71.
82. Vehring R, Lechuga-Ballesteros D, Joshi V, Noga B. Cosuspensions of microcrystals and engineered microparticles for uniform efficient delivery of respiratory therapeutics from pressurized metered dose inhalers. *Langmuir.* 2012;28:15015–23.
83. Ferguson GT, Hickey AJ, Dwivedi S. Co-suspension delivery technology in pressurized metered dose inhalers for multi-drug dosing in the treatment of respiratory disease. *Respir Med.* 2018;134:16–23.
84. Geller DE, Weers J, Heuerding S. Development of an inhaled dry-powder formulation of tobramycin using PulmoSphere™ technology. *J Aerosol Med Pulm Drug Deliv.* 2011;24:175–82.
85. Ramachandran C, Fleisher D. Transdermal delivery of drugs for the treatment of bone disease. *Adv Drug Deliv Rev.* 2000;42:197–223.
86. Keservani RK, Bandopodhyay S, Bandyopahyay N, Sharma AK. Design and fabrication of transdermal/skin drug-delivery system. In: Tekade RK, editor. *Drug Delivery Systems.* New York, NY: Academic Press; 2020. p. 131–78.
87. Despotopoulou D, Lagopati N, Pispas S, Gazouli M, Dematzos C, Pippa N. The technology of transdermal delivery nanosystems: From design and development to preclinical studies. *Int J Pharm.* 2022;611:121290.
88. Flynn GL. Cutaneous and transdermal delivery: Processes and systems of delivery. In: Banker GS, Rhodes CT, editors. *Modern Pharmaceutics.* 3rd ed. New York, NY: Marcel Dekker; 1996. p. 239–98.
89. Foldvari M. Non-invasive administration of drugs through the skin: Challenges in delivery system design. *Pharm Sci Tech Today.* 2000;3:417–25.
90. Hadgraft J. Skin, the final frontier. *Int J Pharm.* 2001;224:1–18.
91. Lauterbach A, Muller-Goymann CC. Applications and limitations of lipid nanoparticles in dermal and transdermal drug delivery via the follicular route. *Eur J Pharm Biopharm.* 2015;97:152–63.
92. Burkoth TL, Bellhouse BJ, Hewson G, Longridge DJ, Muddle AG, Sarphie DF. Transdermal and transmucosal powdered drug delivery. *Crit Rev Ther Drug Carrier Sys.* 1999;16:331–84.
93. Mishra DK, Pandey V, Maheshwsari R, Ghode P. Cutaneous and transdermal drug delivery. In: Tekade RK, editor. *Nasic Fundamentals of Drug Delivery.* Amsterdam, The Netherlands: Elsevier; 2019. p. 595–650.
94. Sahu DK, Ghosh G, Rath G. Introduction to transdermal drug delivery system. In: Chappel E, editor. *Drug Delivery Devices and Therapeutic Systems.* New York, NY: Academic Press; 2021. p. 309–23.
95. Wang R, Bian Q, Xu Y, Xu D, Gao J. Recent advances in mechanical force-assisted transdermal delivery of macromolecular drugs. *Int J Pharm.* 2021;602:120598.
96. Vogt A, Wischke C, Neffe AT, Ma N, Alexiev U, Lendlein A. Nanocarriers for drug delivery into and through the skin. *J Controlled Release.* 2016;242:3–15.
97. Dragicevic H, Maibach H. Combined use of nanocarriers and physical methods for percutaneous penetration enhancement. *Adv Drug Deliv Rev.* 2018;127:58–84.
98. Moffatt K, Donnelly RF. Microneedle technology. In: Chappel E, editor. *Drug Delivery Devices and Therapeutic Systems.* New York, NY: Academic Press 2021. p. 345–66.
99. Lamey PJ, Lewis MAO. Buccal and sublingual delivery of drugs. In: Florence AT, Salole EG, editors. *Routes of Drug Administration.* London, UK: Wright; 1990. p. 30–47.

100. Hua S. Advances in nanoparticulate drug delivery approaches for sublingual and buccal administration. *Frontiers in Pharmacology*. 2019;10:1328.
101. Morales JO, Brayden DJ. Buccal delivery of small molecules and biologics: Of mucoadhesive polymers, films, and nanoparticles. *Current Opinions in Pharmacology*. 2017;36:22–8.
102. Hua S. Physiological and pharmaceutical considerations for rectal drug formulations. *Frontiers in Pharmacology*. 2019;10:1196.
103. Mackay M, Phillips J, Hastewell J. Peptide drug delivery: Colonic and rectal absorption. *Adv Drug Deliv Rev*. 1997;28:253–73.
104. Chien YW. Mucosal drug delivery: Potential routes for non-invasive systemic administration. In: Chien YW, editor. *Novel Drug Delivery Systems*. New York, NY: Marcel Dekker; 1992. p. 139–96.
105. Edwards C. Physiology of the colorectal barrier. *Adv Drug Deliv Rev*. 1997;28:173–90.
106. Alexander N, Baker E, Kaptein M, Karck U, Miller L, Zampaglione E. Why consider vaginal drug administration? *Fertil Steril*. 2004;82:1–12.
107. Hussain A, Ahsan F. The vagina as a route for systemic drug delivery. *J Controlled Rel*. 2005;103:301–13.
108. Srikrishna S, Carozo L. The vagina as a route for drug delivery: A review. *Int Urogynecol J*. 2013;24:537–43.
109. Osmalek T, Froelich A, Jadach B, Tatarek A, Gadzinski P, Falana A, et al. Recent advances in polymer-based vaginal drug delivery systems. *Pharmaceutics*. 2021;13:884.
110. Choudhary A, Das S, Kar M. A review of novelty and potentiality of vaginal drug delivery. *Int J Pharm Tech Res*. 2011;3:1033–44.
111. Baloglu E, Senyigit ZA, Karavana SY, Bernkop-Schnurch A. Strategies to prolong the intravaginal residence time of drug delivery systems. *J Pharm Pharmaceut Sci*. 2009;12:312–36.
112. Carmella C, Rossi S, Ferrari F, Bonferoni MC, Sandri G. Mucoadhesive and thermogelling systems for vaginal drug delivery. *Adv Drug Deliv Rev*. 2015;92:39–52.
113. dosSantos AM, Goncalves-Carvalho S, Araujo HHS, Carvalho GC, Greniao MPD, Chorilli M. Recent advances in hydrogels as strategy for drug delivery to vaginal infections. *Int J Pharm*. 2020;590:119867.
114. dasNeves J, Nunes R, Machado A, Sarmento B. Polymer-based nanocarriers for vaginal drug delivery. *Adv Drug Deliv Rev*. 2015;92:53–70.
115. Subi MTM, Selvasudha N, Vasanthi HR. Vaginal drug delivery system: A promising route of drug administration for local and systemic diseases. *Drug Disc Today*. 2024;29:1–7.
116. Lee VH, Robinson JR, Lee VHI. Influence of drug properties and routes of administration on the design of sustained and controlled release systems. In: Robinson JR, Lee VHI, editors. *Controlled Drug Delivery: Fundamentals and Applications*. New York, NY: Marcel Dekker; 1987. p. 3–94.
117. Notario-Perez F, Cazorla-Luna R, Martin-Illana A, Galante J, Ruiz-Caro R, dasNeves J, et al. Design, fabrication and characterization of drug-loaded vaginal films: State-of-the-art. *J Controlled Rel*. 2020;327:477–99.
118. Antimisiaris SG, Mourtas S. Recent advances on anti-HIV vaginal delivery systems development. *Adv Drug Deliv Rev*. 2015;92:123–45.
119. Malcolm RK, Boyd PJ, McCoy CF, Murphy DJ. Microbicide vaginal rings: Technological challenges and clinical development. *Adv Drug Deliv Rev*. 2016;103:33–56.

10 General Conclusions

From the earliest use of powders alone to their use in capsules and tablets, the method of manufacture, physical form, and characteristics of a dosage form have been known to play key roles in the performance and ultimately the efficacy of drugs. Significant advances in the last 50 years in targeted systems to control drug delivery to the site of action or absorption have improved efficacy while minimizing toxicity. The therapeutic effect of drugs delivered in this manner, as compared to traditional dosage forms, is more highly dependent on the nature of the particulate system being employed, and sophisticated methods of analysis are required for its characterization. This book outlines the full complexity of pharmaceutical particulate systems and makes the case for giving consideration to all aspects of their performance, from manufacturing processes to therapeutic effect.

A classification framework is required in which to discuss the form of a particle before manufacturing can begin. The field of solid-state physics has already built the language by which particle structure can be discussed. The core element of this language is the crystal system defining the molecular array within a particle and the crystal habit, which describes the outward appearance of the particle following growth from a supersaturated medium. Particles can be manufactured in a multitude of ways, but each one of them will result to a greater or lesser degree in crystalline or amorphous particles. The form of the particles will dictate their physicochemical properties, which, in turn, through forces of interaction will dictate the state of the powder of which they are components.

Once particles have been constructed, and assessed as geometric forms, their dimensions must be described in general terms. Particle size and distribution form the basis for such descriptors. Particle size can be considered in terms of individual particles for which, again, common expressions must be employed, such as equivalent volume, surface, or projected area spheres. However, particles are found in distributions of sizes around a mean value. These distributions may be fitted to statistical distributions or mathematical functions, but it is important to recognize that such data models are only approximations of the real distribution. Therefore, the raw data may ultimately be the most valuable description of the size distribution. Particle size and distribution data may be employed to predict or understand phenomena ranging from dissolution properties to blending characteristics to bioavailability of the product.

Clearly, particle size and distribution are key properties from which many performance characteristics of a dosage form can be derived. However, before such correlations can occur, it is essential to select or devise an appropriate particle sizing technique. Of the range of methods available, some are direct, such as microscopy, whereas others are indirect, such as light scattering. It is imperative that the user understand the underlying theory for a method. Moreover, this understanding should

DOI: 10.1201/9781003470151-10

be the basis for selecting appropriate methods for the application of the data. For example, Stokes' diameter, based on sedimentation, is most appropriate for suspension formulations. The complementarity of particle sizing techniques should never be understated. Because each particle size descriptor and its method of analysis reflect a different facet of the same particle or population of particles, a great opportunity exists to use a multitude of techniques to build a picture of the powder.

At this juncture, the potential for error may seem to have been obviated by considering all of the issues above. Unfortunately, in routine activities, many scientists erroneously come to this conclusion. The accuracy of any particle size determination depends most significantly on sampling. The entire population of particles can rarely be used as the sample, so the issue of representative sampling becomes very important. This sampling may be an essential part of whatever particle sizing technique is employed, or it may be incumbent on the scientist to sample the powder prior to analysis and use. Fortunately, sampling has been studied in great detail and the statistical bases for good practices in sampling are established. Consequently, adopting good laboratory or manufacturing procedures based on sound engineering principles should allow the most accurate and precise data to be obtained.

The manufacture of the particle itself has been considered, but ultimately this will become a discrete element in the dosage form. It is frequently the case that drug particles are mixed with excipient particles to form homogeneous dispersions for a variety of purposes. It may then be necessary to examine the particulate properties of the blend, which in turn will require adequate sampling procedures.

Finally, the implications of the method of manufacture and characteristics must be considered from the perspective of in vitro testing for the quality of the product and in vivo performance for the efficacy and toxicity of the product. The fondest wish of pharmaceutical scientists is to correlate in vitro with in vivo performance and to approach the metaphorical Holy Grail of an in vitro–in vivo correlation (IVIVC). Although a great deal of research has been conducted in this area with some degree of success, it remains true that in vitro testing is, at best, a qualitative predictor of in vivo performance. Consequently, these issues must be considered separately. The in vitro performance of a fully characterized product may be considered in terms of a number of parameters of varying importance. For example, flow is important for vial filling, dissolution is important for oral absorption, and dispersion is important for the delivery of aerosols. Examples of in vivo properties include lung deposition for aerosol products and oral bioavailability (which relates to in vivo dissolution).

Pharmaceutical scientists now have many techniques at their disposal that allow them to appreciate pharmaceutical particulate dispersed systems with the clarity once felt for visualizing sand, gravel, stones, rocks, and boulders. Increasingly, particle characterization is recognized as a key to the quality and therapeutic performance of particulate medicines. The importance of particle properties in the action of drugs delivered from sophisticated targeted and controlled delivery systems has increased the rigor with which they are evaluated. Discussions of the benefits of crystal engineering are becoming more common, and the control of processes for the production of particles is of interest to regulatory and pharmacopeial groups. The complementarity of particle sizing techniques in elucidating the nature of particle morphology may be equated to the allegory of the blind ones and the elephant. Depending on which feature is measured, different conclusions as to the nature of the object might be drawn. Therefore, the complementarity of techniques gives greater insight into the entire nature of the particle or powder. Nevertheless, some unfounded consternation remains on the part of novices that all techniques do not give the same result.

Integrating optimal sampling techniques and particle sizing methods with their process application by considering the nature of particle interactions in the context of the environment in which they exist forms the basis for good practice in pharmaceutical particulate science. By educating those new to these concepts, and embracing the experiences of other industries with respect to particulate matter, the prospects of increased understanding, improved medications (in terms of both quality and efficacy), and appropriate regulations seem assured.

Figure 10.1 illustrates the core functions in pharmaceutical particulate science and their relationship to materials in the molecular or solid state. Once a drug molecule has been discovered or isolated, it will be evaluated in a preliminary fashion for efficacy. Assuming that the potential has been established for the drug to be active in specific diseases, the drug is subjected to a manufacturing process to prepare particles. These particles can then be characterized to establish specifications for the quality of the final dosage form. In vitro and in vivo performance can be evaluated to establish the potential for efficacy in the final product (note that some of the in vitro performance characteristics are used to establish product quality). The cycle is then complete because efficacy is the justification for initiating the product development.

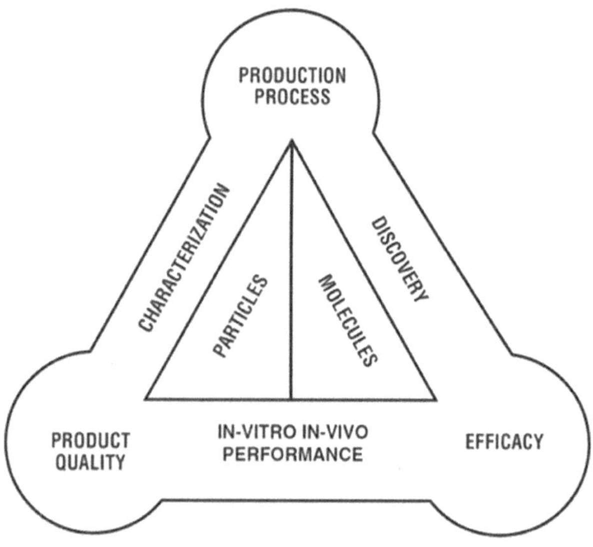

FIGURE 10.1 Core functions in pharmaceutical particulate science.

Clearly, particulate science plays a major role in the therapeutic application, and systematic studies are essential to the quality of the product. A greater understanding of all aspects of particle formation, morphology, degree of subdivision, and behavior will lead to increasingly sophisticated and therapeutically viable dosage forms that will ultimately revolutionize the delivery of drugs.

Index

FSC
www.fsc.org
MIX
Papier | Fördert gute Waldnutzung
FSC® C083411

Kolibri 360 GmbH
Ferdinand-Jühlke-Straße 7
99095 Erfurt, Deutschland
produktsicherheit@kolibri360.de